Microbial Life History

T0074535

Microbial Life History

The Fundamental Forces of Biological Design

STEVEN A. FRANK

Princeton University Press

Princeton and Oxford

Published by Princeton University Press
41 William Street, Princeton, New Jersey 08540
99 Banbury Road, Oxford OX2 6JX

press.princeton.edu

Library of Congress Cataloging-in-Publication Data
Names: Frank, Steven A., 1957– author.
Title: Microbial life history : the fundamental forces of
    biological design / Steven A. Frank.
Description: Princeton : Princeton University Press, [2022] |
    Includes bibliographical references and index.
Identifiers: LCCN 2021059577 (print) | LCCN 2021059578
    (ebook) | ISBN 9780691231198 (paperback) | ISBN
    9780691231204 (hardback) | ISBN 9780691231181 (ebook)
Subjects: LCSH: Microbial metabolism–Evolution.
Classification: LCC QR88 .F73 2022 (print) | LCC QR88 (ebook) |
    DDC 576.8/2-dc23/eng/20220120
LC record available at https://lccn.loc.gov/2021059577
LC ebook record available at https://lccn.loc.gov/2021059578

British Library Cataloging-in-Publication Data is available

Editorial: Alison Kalett and Hallie Schaeffer
Production Editorial: Mark Bellis
Cover Design: Heather Hansen
Production: Lauren Reese
Publicity: Matthew Taylor and Charlotte Coyne
Copyeditor: Cyd Westmoreland

Typeset by the author in TEX
Composed in Lucida Bright

10 9 8 7 6 5 4 3 2 1

E. M. Forster ... set out the difference between a story and plot. "The king died and then the queen died" is a story, he wrote. But a sense of causality is needed to make a plot more than just a sequence of events. "The king died and then the queen died of grief" is a plot.

—*The Economist*[408]

Contents

Preface

Microbes vary. Some grow quickly, using resources inefficiently. Others grow slowly, achieving efficient reproductive yield.

Why do evolutionary processes lead to such diversity? To answer that question, we must ask: How do the fundamental evolutionary forces shape biological design?

For example, comparing scarce versus abundant food, how do we expect evolutionary forces to alter growth rate and metabolic design?

Comparison provides the key. If we can predict how traits change when comparing different conditions, then we can reasonably say that we understand the fundamental evolutionary forces of design.

We face two challenges. Conceptually, we must understand the fundamental forces to make good comparative predictions. Empirically, we must translate data into the weight of evidence for or against the causal role of specific forces.

This book develops comparative predictions for microbial traits. Recent advances in microbial studies provide an ideal opportunity to test those predictions about diversity and design, perhaps the greatest problems in biology.

I received financial support from the Donald Bren Foundation, the US National Science Foundation, and the US Army Research Office.

A book is nothing without a home and someone who believes in you. Thank you, Robin.

1 Microbial Design

In the past, changes in gene expression and metabolic strategies across growth conditions have often been attributed to the
optimization of ... growth rates. However, mounting evidence
suggests that cells are capable of significantly faster growth
rates in many conditions. ... Based on these observations, it
is clear that [design] objectives other than optimization of ...
growth rates must be considered to explain these phenotypes.

—Markus Basan[27]

Why don't microbial cells grow as fast as possible? Perhaps cells trade
growth rate for other attributes of success.

One widely discussed tradeoff concerns rate versus yield. Growing
faster uses resources inefficiently. Resources wasted to increase metabolic rate lower the resources available to build new biomass. Fast
growth rate reduces the reproductive yield.[317,444]

Suppose we observe microbes that grow more slowly than the maximum rate that they could achieve. We see mutations that enhance
growth. How can we know if the tradeoff between growth rate and yield
dominates in metabolic design?

Typically, we cannot know. An observed rise in rate and decline in
yield supports the tradeoff. But rejecting the rate-yield tradeoff hypothesis is difficult. For example, the microbes may produce toxins to kill
competitors. If competitors are absent in our study, we may see increases
in both rate and yield as the unobserved toxin production declines.

Other tradeoffs may be hidden. Perhaps growth trades off with dispersal. Maybe the microbes typically grow under iron-limited conditions
and must trade growth rate for scavenging iron.

We could measure more tradeoffs. Although helpful, that approach
ultimately fails. We can never estimate the many tradeoffs across the
full range of natural conditions that shaped design.

Given those difficulties, how can we understand why growth rate is sometimes maximized and other times not? In general, how can we understand the forces that shape the design of microbial traits, such as dispersal, resource acquisition, defense, and survival?

I advocate comparative hypotheses. As a focal parameter changes, we predict the direction of change in a trait. For example, as the genetic heterogeneity among competitors rises, we predict an increase in growth rate.[130,317] If the predicted direction of change tends to occur, then the focal parameter associates with a causal force that shapes the trait, revealing the fundamental forces of biological design.

This book divides into two parts. The first part presents the conceptual tools for making comparative predictions. The second part develops comparative predictions for metabolic traits.

We can use this approach to make comparative predictions for the full range of microbial traits, providing a general method for the study of biological design.

1.1   How to Read

Part 1 sets the theoretical background. How does one form and test predictions about the forces that shape biological design?

Part 2 turns to unsolved puzzles in microbial metabolism. How can we use Part 1's principles for the study of design to advance the understanding of microbial evolution?

Readers primarily interested in microbes may wish to start with the second part. As particular concepts arise in that second part, one may follow the pointers to the first part to fill in the background.

Readers primarily interested in evolutionary concepts may wish to start with the first part. The second part illustrates how to turn those concepts into a fully realized program of empirical study.

Although each part stands alone, the real value comes from the synergy between parts. Full progress demands combining Part 1's evolutionary concepts and general principles for studying causality with Part 2's application to metabolism, the engine of life.

That pairing between theory and application provides the best way to study the forces that have shaped biological design.

To help readers find their preferred starting point and path through the book, the following sections briefly summarize each chapter.

1.2   Theoretical Background

Organismal traits often seem designed to solve environmental challenges. Presumably, natural processes have shaped design. However, the underlying processes can be difficult to observe.

How can we study those causal forces of design? Somehow, we must link the hidden forces to the observed traits. Part 1 develops the theoretical background to meet that challenge.

Chapter 2 defines design in relation to biological fitness, the ultimate measure of success. Three fundamental forces of design often dominate. Marginal values measure trading one design for another. Reproductive values weight different components of fitness, such as reproduction, survival, and dispersal. Generalized kin selection links the similarity of interacting individuals with the transmission of traits through time.

Chapter 3 turns to the causal analysis of design. We can rarely match organismal traits to the forces of design that shaped those traits. Many particular forces played a role. We cannot measure or infer all of them.

Instead, we must focus on change. Can we predict how change in a specific factor alters a particular trait? For example, how does increasing genetic variability between competitors alter reproductive rate?

Comparing states of a particular factor isolates partial causality, the change caused holding all else constant. Comparative prediction becomes the building block of causal understanding. How does a changed factor alter a trait, mediated by a fundamental force of design?

Chapter 4 illustrates comparative predictions. The examples link changes in environmental factors to predicted changes in the metabolic traits of microbes. Each hypothesis associates the predicted change in a metabolic trait to a causal force of biological design.

The following chapters of Part 1 fill in the theoretical background needed to develop comparative predictions. Part 2 uses that theory to make comparative predictions about organismal design, with emphasis on microbial metabolism.

Chapter 5 reviews various forces that shape biological design. Marginal values, reproductive values, and generalized kin selection play key roles, as noted above. Natural history modulates forces of design. Examples include demography and complex life cycles, the scaling of spatial and temporal environmental variability, and the different timescales over which competing design forces act.

Chapter 6 notes that biological design concerns organismal traits. However, the nature of traits often remains vague. Different problems arise when studying the evolutionary origin of traits versus the modification of traits. Some traits change within an organism in response to the environment. Other traits may be genetically fixed, varying only between individuals rather than within them.

Chapter 7 extends discussion of traits that vary within an individual. Much of evolutionary design concerns the control of such traits in response to environmental signals. This chapter reviews principles of engineering control theory as they may be applied to biological design. Error-correcting feedback is perhaps the single greatest principle of design in both human-engineered and biological systems.

Chapter 8 contrasts this book's comparative predictions with historical antecedents. Darwin developed comparison in the study of adaptation. Classic phylogenetic comparative methods extended Darwin's vision.

This book differs primarily in the scale of change. Prior analyses typically studied change between species or higher taxa. By contrast, design forces often act at smaller scales of change. Those smaller scales set the focal point for this part's theory and the following part's application to microbial metabolism.

1.3   The Design of Metabolism

In microbes, large populations and short generation times provide opportunity to observe small-scale changes in action. Progress in technology and measurement opens new windows onto those small-scale changes. Part 2 takes advantage of this new era in the study of biological design to advance the testing of comparative hypotheses.

Chapter 9 explains the focus on metabolism. Extracting and using the free energy driving force from food is a universal challenge of life. Microbial metabolism provides a good starting problem to sharpen our tools in the study of biological design.

Chapter 10 illustrates comparative hypotheses and tests by analyzing microbial growth rate, typically measured as the increase in biomass. Growth rate seemingly provides the simplest trait by which to measure fitness, the long-term contribution to the future population.

However, tradeoffs arise. Faster short-term growth may use resources inefficiently. Lower efficiency reduces reproductive yield per unit food uptake, slowing long-term growth as food gets used up. Comparatively, decreasing the available food raises the marginal gains for yield efficiency. Enhanced gains for yield predict lower short-term growth rate, driven by the fundamental force of marginal valuations between alternatives.

This chapter lists many comparative hypotheses. Those hypotheses link changes in natural history to predicted changes in growth rate. The analysis then turns to testing comparative hypotheses. Examples illustrate the kinds of data that have recently been collected in natural and laboratory populations.

Chapter 11 develops the universal challenge of extracting free energy from food to drive the processes of life. The thermodynamic driving force of free energy comes from moving low entropy electrons in food to high entropy electrons in final electron acceptors, such as oxygen.

Metabolic design exploits the increasing entropy between food and final electron acceptors to drive coupled reactions that decrease entropy. The decreased entropy of the driven reactions creates the ordered molecules of life or the entropy disequilibria, such as ATP versus ADP, that act as storage batteries to drive subsequent order-creating processes.

Textbook descriptions of biochemical thermodynamics often fail to emphasize how the entropy disequilibria in food drive the entropy disequilibria of life.[18,47,294] Studying metabolic design requires focus on the flux of those coupled disequilibria through metabolic cascades.

Metabolic flux also depends on the resistance to reactions from chemical activation barriers. Cells modulate resistance by using enzyme catalysts or by changing the biochemical conditions. Net flux depends on the thermodynamic driving force divided by the resistance to reaction, an analogy with Ohm's law of electric current flow.

Chapter 12 describes how cells modulate flux by altering the thermodynamic driving force. The greater the displacement of a reaction from equilibrium, the greater the driving force and the rate of reaction. High driving force also causes the loss of potentially usable entropy change, typically dissipated as heat.

This chapter analyzes the design of glycolysis in terms of the thermodynamic tradeoff between reaction rate and usable entropy yield. Recent technical advances allow direct in vivo measurement of the driving force for individual reactions within the glycolytic cascade.

Those direct measurements open up new possibilities to study comparative hypotheses. For example, environmental changes in cellular competition and genetic variability may alter the fine-scale design of metabolic flux control. Large-scale biochemical changes between alternative glycolytic pathways also pose interesting puzzles of design.

Overflow metabolism presents a key challenge. Many microbes excrete post-glycolytic products that contain most of the usable entropy in the original food source. Why overflow usable food? Disequilibria, thermodynamic driving force, and the tradeoff between rate and yield play important roles. Changed conditions alter overflow, providing a model to test comparative hypotheses about metabolic design.

Chapter 13 discusses the modulation of flux by altering the resistance of reactions. Mechanisms include varying enzyme concentration, modifying enzyme structure, and spatially separating reactants.

Changes in metabolic design may alter thermodynamic driving force or the resistance to reactions. Small changes typically occur by modulating current biochemical pathways. Larger changes may lead to different biochemical pathways. Other design goals shape pathways, such as the need for precursors to build particular molecules.

Constraining forces interact with design forces. For example, cell size constrains space for protein catalysts. Limited proteins impose tradeoffs between the potential to modulate different reactions.

Flux control has been widely discussed. However, clearly specified comparative hypotheses remain scarce with regard to the forces of design and constraint that have shaped metabolic diversity. This book sets the foundation on which to build comparative hypotheses and provides many examples of such hypotheses.

Chapter 14 turns to the observed diversity in metabolic pathways. The biochemical detail in this chapter raises many puzzles, setting a challenge for comparative predictions and tests of metabolic design.

In one example, different glycolytic pathways have different yields of ATP, NADH, and NADPH, each of which create distinct disequilibria that drive different cellular processes. In another example, the diverse final electron acceptors of catabolism create different entropy gradients, which greatly influence metabolic design. Weak gradients pose special design challenges.

Metabolic electron flow sometimes happens between cells of the same or different species. Distributed electron gradients raise novel puzzles in metabolic design. Those puzzles often depend on how particular biochemical disequilibria enhance or limit electron flow.

This chapter also analyzes the regulation of alternative sugar catabolism within cells and cellular shifts between different complex carbohydrate food sources. The chapter's conclusions synthesize puzzles of design for variant pathways.

Chapter 15 emphasizes tradeoffs, which set the basis for design. For example, faster growth reduces food use efficiency. Less permeable membranes protect against attack but slow resource uptake.

However, particular tradeoffs often fail to reveal design. Suppose growth rate, yield efficiency, and defense trade off. Less attack reduces investment in defense, potentially increasing both growth rate and yield. Without measurement of defense, one might see only the simultaneous rise in rate and yield, apparently contradicting the rate-yield tradeoff.

Comparative hypotheses about the tradeoffs themselves may help. For example, more abundant food weakens the tradeoff between growth rate and yield efficiency.

The more completely one understands the range of potential tradeoffs, the more effectively one can make comparative predictions. This chapter provides a preliminary catalog of the tradeoffs that shape the metabolic design of microbes.

Chapter 16 highlights the forces that shape overflow metabolism, the cellular excretion of usable food. Several challenges for inferring design emerge. Forces act over different timescales. Each empirical method reveals particular forces and timescales while hiding others.

Progress requires explicit consideration of the challenges and limitations in the study of biological design. The importance of clear comparative predictions and partial causation rises once again.

Chapter 17 continues the analysis of model problems in metabolic design. Part 1's forces of design play an important role as we broaden the range of metabolic traits and natural history.

When exposed to multiple foods, how do cells express alternative catabolic pathways? Sometimes, preferred foods repress pathways for other foods. Other times, cells simultaneously express different pathways. In some clonal populations, cells differ in expression patterns. Various design forces shape expression. Testable comparative predictions follow.

How do cells overcome limited access to final catabolic electron acceptors such as oxygen? Cable bacteria form filaments with electric wires. The wires pass electrons from anoxic zones to oxic zones, creating strong catabolic flux. Linked cells form various multicellular lengths, altering life cycles, spatial competition, and the forces of design.

Other species use extracellular shuttle molecules to move electrons from cell surfaces to distant electron sinks. Shuttles, once released from producing cells, can be used by any neighboring cells. Such publicly shareable resources create special challenges. Demography and genetic mixing alter design forces in predictable ways.

When life cycles pass through habitats that prevent catabolism, how do cells store and use resources? Microbial wastewater treatment provides an interesting model system. The treatment passes bacteria through alternate anaerobic and aerobic habitats. Food is available only during the anaerobic phase. However, lack of oxygen prevents catabolism.

In that anaerobic habitat, cells transform food into internal storage. During the aerobic phase, cells catabolize the internal stores. Varying the alternative habitats changes the demographic forces of design.

Wastewater treatment and other industrial applications provide excellent model systems to test comparative predictions about the forces that shape metabolic design.

Chapter 18 revisits problems in the study of biological design.

Part 1

Theoretical Background

2 Forces of Design

Are the plants not perhaps the real adherents of the doctrine of marginal utility, which seems to be too subtle for man to live up to?

—R. A. Fisher, Letter to Leonard Darwin[36]

Natural selection favors traits that increase success. To start, we must understand what we mean by *success*.

The first section discusses fitness, the ultimate biological measure of success. Three fundamental forces influence fitness: kin selection, reproductive value, and marginal value.[122] Each force expresses how changed traits drive change in the future genetic composition of the population.

The second section considers difficulties in measuring fitness. One can rarely measure all components of success. Progress requires an indirect and informative way of gaining insight into the forces that shape design. The following chapter promotes comparative analysis, perhaps the only realistic solution.

2.1 What Is Fitness?

Fitness is the genetic contribution to the future population, the ultimate measure of success.[106] In the study of design, we ask whether a changed trait increases or decreases future genetic contribution. Altered traits that enhance genetic contribution spread. Altered traits that reduce genetic contribution disappear.

Tradeoffs occur. Is it better to reproduce faster but ultimately make fewer progeny? Or does natural selection favor slower reproductive rate and ultimately more total progeny? What about other tradeoffs with reproductive rate, such as survival or the ability to disperse and colonize new locations?[392]

A full measure of overall success requires that we combine the different components, such as reproduction, survival, and dispersal, to obtain a proper measure of fitness.

The best way to appreciate the multifaceted nature of fitness is to analyze the forces that shape various microbial traits. Part 2 provides many examples. Here, I highlight general principles to pave the way for later applications. Chapter 5 develops the theory in more detail.

THREE FUNDAMENTAL FORCES

Kin selection and similarity selection.—The genetic and phenotypic similarities between neighboring individuals influence genetic contribution to the future population. For example, an individual that outcompetes a genetically identical clonemate adds little to its ultimate genetic representation in the future population. By contrast, an individual that outcompetes a genetically distinct competitor enhances its future representation in the gene pool.[122,167]

Similarities between organisms often arise by kinship. However, other causes of similarity occur. For example, organisms may sort themselves spatially based on their particular traits.[448] Or synergistic traits between organisms may increase their spatial association by enhancing the joint survival of successful pairs.[113,119] Processes that enhance or degrade spatial associations can influence similarity more strongly than kinship.

Genetic similarity can influence potentially cooperative traits. For example, secreted siderophores for iron uptake or secreted enzymes for exodigestion can be publicly shareable goods that enhance the growth rate of neighbors.[441] If those neighbors are genetically similar to the secreting individual, then the cooperative benefits to neighbors can enhance the secretor's genetic representation in the future population.

Those competitive and cooperative social aspects of fitness are relatively easy to study. By measuring the correlations between interacting individuals and basic fitness components, such as reproductive rate and yield, one obtains a reasonable estimate of the relation between traits and fitness. Simple theories about social traits can be tested directly, particularly when measuring genetic correlations, reproductive rates, and total reproductive yield under controlled laboratory conditions.

Demography and reproductive value.—Simple concepts of fitness based on similarity typically ignore essential demographic aspects of populations. Demography includes the intrinsic aging of resource patches, the variation in resource quality over space, and the key roles of dispersal and successful colonization in determining the long-term genetic contribution to the future population.

Those demographic aspects lead to reproductive value, the second force that contributes to a total measure for fitness. Reproductive value describes the relative strength of each fitness component with regard to its contribution to the future genetic composition of the population.[61,106] For example, in a growing population, faster reproduction is better than greater survival because offspring in a growing population form an expanding clonal lineage. In a declining population, greater survival is typically better than faster reproduction because offspring in a declining population form a shrinking clonal lineage.

The relative valuation of reproduction versus dispersal also requires a proper translation into future contributions. In a rich and uncrowded habitat, a nondispersing offspring has a relatively large expectation of contribution to the future population. Rapid reproduction and low dispersal may be favored. In a poor and crowded habitat, a nondispersing offspring has a relatively low expectation of future contribution. Slow reproduction and high dispersal may be favored.

In general, one cannot simply count up the individuals that result from survival, reproduction, and dispersal to obtain a measure of fitness. Instead, each component of success must be translated into fitness by the two key forces.[122,405] Kin or similarity selection, in the context of competition and cooperation, determines how changes in traits alter genetic contributions to the population. When calculating how changes in traits alter total genetic contributions to the future population, reproductive value determines how to weight different components of fitness.

Marginal value.—The third force compares gains and losses of different fitness components on a common scale. Suppose, for example, that microbial growth rate trades off with yield. To evaluate how natural selection shapes metabolism, we calculate how much additional growth rate can be achieved for each small (marginal) loss in yield. A small reallocation of resources from yield to growth defines the marginal costs and benefits.

With an excess of available resources, large marginal gains in short-term growth rate may impose relatively small marginal losses in long-term reproduction because the wasted resources to fuel faster growth can be offset by the excess supply.

By contrast, with limited resources, faster growth may deplete resources sooner. Depleted resources cause greater marginal losses in long-term reproductive yield.

In general, marginal valuations provide a common currency with which to analyze tradeoffs.

UNITS OF SELECTION AND TIMESCALE

The three forces of fitness define how traits, such as metabolism, influence the contribution of genes to future generations. That description of forces leaves open the question of which genes. For example, a trait may have different fitness consequences for horizontally transmitted genes on plasmids and vertically transmitted genes on chromosomes. Such conflicts between different genetic units of selection can powerfully influence the evolution of traits.[54,205] In this book, I typically focus on a simple notion of chromosomal (vertical) success, unless otherwise noted.

I have defined fitness in terms of genetic contribution to future generations. However, selection may work differently on different timescales, associated with different periods in the future.[133,236] Suppose, for example, that a relatively slow growth and high yield metabolism provides the greatest contribution of genes to the distant future. By contrast, a mutant with relatively rapid growth and lower yield increases immediately, despite having lower long-term success. The long and short timescales conflict. That conflict may lead to heterogeneity in the tuning of metabolism.

The relative dominance of the different timescales depends on various factors. For example, local interactions over short timescales may favor rapid growth to outcompete neighbors. By contrast, distant interactions over longer timescales may favor slow growth and greater reproductive yield to outperform remote groups when competing by dispersal for colonization of new resource patches.

We can develop comparative predictions for how changes in environmental attributes and demographic parameters alter the balance of short and long timescales and the tuning of microbial traits. Later chapters present many comparative predictions.

2.2   The Difficulty of Measuring Fitness

The demographic components of reproductive value illustrate the challenges of empirical study. To measure long-term genetic contribution, one must evaluate success over the complete cycle of growth in a resource patch and colonization of new resource patches. Measuring success over a complete cycle may not be easy to do when the stages of growth in a particular location are complex and resource patches vary over time and space.

Nonetheless, the analysis of microbial design must confront the full measurement of fitness. It is often misleading to focus on a single component, such as a short-term measure of growth, or on a single tradeoff, such as survival versus dispersal.

Comparison solves the problem of measuring fitness. Before turning to comparison in the next chapter, let us first consider more fully the difficulties of testing hypotheses about design.

The study of parasite virulence provides an interesting historical example. The early theory began with a few key tradeoffs, such as virulence versus transmission or, equivalently, survival within hosts versus dispersal to new hosts.[14,52,117] Within a few years, the theory developed a broader synthesis that fully combined the concepts of kin selection and reproductive value into a comprehensive understanding of fitness.[122] Yet, despite many thoughtful developments of the theory, the dominant slogan of empirical study has often been reduced to the tradeoff between virulence and transmission, as if nothing else mattered.

Empirical studies of parasites sometimes fail to find clear evidence of a tradeoff between virulence and transmission.[1,175] From that failure, one might conclude that the theory cannot explain the design of parasite traits in relation to virulence by using the fundamental concepts of fitness and adaptation. However, the real problem is that any attempt to focus on a single tradeoff or a single dimension of fitness will always yield inconsistent results and an apparent failure of the evolutionary principles of biological design.

For example, in an expanding epidemic, enhanced transmission and dispersal to new hosts have a stronger reproductive value weighting because a growing population corresponds to an expanding descendant lineage. By contrast, in a declining epidemic, reduced virulence and greater within-host survival have a stronger reproductive value weighting because a declining epidemic corresponds to a shrinking lineage of descendants associated with transmission to new hosts.[117,230] Aggregating over different epidemic patterns may lead to inconsistent virulence-transmission tradeoffs.

The study of microbial design is at risk of a similar failure. The proper measure of fitness is conceptually challenging and empirically difficult. Faced with those difficulties, it is natural that people have sought simple aspects of success, such as growth rate or relative dominance in pairwise competitions. Those simple attributes can be measured precisely. But precision in limited dimensions does not substitute for full analysis.

Even a strong attempt at full analysis will probably fail. For example, suppose we find a microbe that grows at a rate far below its potential maximum. Numerous mutations increase growth rate. What is the function of growing slowly?

Ideally, we would measure all of the different components of fitness and all of the tradeoffs between those components. Maybe, under severe resource stress, the slow-growing design survives better than a fast-growing alternative. If so, we would then have to consider how often the organism faces severe resource stress over time and space.

Would such temporal and spatial stresses be sufficient to claim that the slow-growing type has higher fitness than a fast-growing alternative? If so, why does the microbe not adjust its growth rate to match the conditions, growing more slowly when stress is likely and faster when abundant resources are likely? What about the tradeoff between growth and dispersal under different resource conditions?

The point is that one cannot realistically explain any single phenotype in a particular biological scenario. Very many parameters influence the fitness of that phenotype. One cannot know all of them. If the full measurement of fitness is difficult, what can be done realistically to advance the study of biological design?

3 Comparison and Causality

[Economics] undertakes to study the effects which will be produced by certain causes, not absolutely, but subject to the condition that *other things are equal,* and the causes are able to work out their effects undisturbed. Almost every scientific doctrine ... will be found to contain some proviso to the effect that other things are equal: the action of the causes in question is supposed to be isolated; certain effects are attributed to them, but only *on the hypothesis* that no cause is permitted to enter except those distinctly allowed for.

—Alfred Marshall[262]

The prior chapter showed that one never fully measures fitness. The first section of this chapter argues that comparative predictions provide a way forward. A comparative prediction describes how a change in some condition alters a trait, mediated by a fundamental force of design. The logic of comparative predictions and the broad listing of predictions for microbial traits set the primary themes of this book.

The second section contrasts evolutionary and organismal responses. A changed condition may favor an evolutionary response in the population. A change may also trigger a phenotypically plastic organismal response in individuals. In simple cases, theory makes the same comparative prediction for evolutionary response and organismal response.

The third section develops the notion of a fundamental force as a partial cause. To give a physical example, gravity is a fundamental force that acts as a partial cause of motion but is rarely by itself a complete explanation of motion. Comparative predictions isolate fundamental forces and partial causes.

The fourth section briefly reviews recent progress in causal inference. I set the causal study of organismal design within the larger context of formulating and testing causal hypotheses.

The fifth section specifies the structure of comparative predictions. That section also presents notation for writing comparative predictions.

The sixth section reviews the goals and approach. Subsequent chapters develop comparative predictions. Those predictions reveal the fundamental forces that shape microbial design.

3.1 Comparative Predictions

Testable hypotheses follow from comparative predictions. For example, as the genetic correlation and kin selection relatedness between individuals rise within groups, theory predicts greater cooperative and less competitive trait expression. Tests on microbes support that comparative prediction.[441] Here, I focus on the structure of comparative predictions and the analysis of causality.

ADVANTAGE OF COMPARISON

Microbes often secrete publicly shared factors, such as iron-scavenging siderophores.[216,235] Once released from a cell, the publicly available molecules can be used by neighboring cells. Greater production by an individual cell cooperatively benefits the local group. By contrast, "cheating" nonproducers outcompete neighbors by using the secreted factors of others and saving the cost of production.

Numerous studies support the comparative kin selection prediction in the introductory paragraph of this section. Under high relatedness, relatively more individuals cooperatively produce a shared public good. Under low relatedness, relatively fewer individuals produce the public good, acting as nonproducing cheaters that outcompete neighbors.[216]

In such comparisons, one does not have to measure all components of fitness. The prediction only requires that, aggregated over a variety of common conditions, there be an overall tendency for increased relatedness to associate with increased cooperative trait expression.

NOTION OF CAUSALITY

Comparison provides a reasonable notion of causality. If I repeatedly observe or make a change in condition A, and the predicted direction of change in B tends to happen under a variety of circumstances, then A seems to be a cause of B. We do not have to know all of the factors involved and all of the different conditions. We only need to know that we predicted a particular direction of change, and we tended to see that direction of change.

Of course, confounding correlations and other difficulties can complicate causal inference. Later sections discuss ways to increase the accuracy of causal analysis.

INFERENCE

Statistical inference for comparisons is often simple.[107] If I predict the direction of change in five independent tests, then the probability that I would be right by chance in all five cases is $p = 1/2^5 \approx 0.03$.

The probability that I would be right by chance 59 or more times in 100 trials is also $p \approx 0.03$, from the binomial probability distribution. A prediction with a small tendency in the right direction can provide a significant indication of an underlying force.

We can restate the issue for comparison and public goods. It is difficult to say, in any particular example, whether a certain level of relatedness should be associated with a particular level of public goods secretion. By contrast, we can make a strong prediction about the direction of change in public goods secretion with a change in relatedness.

A comparative directional prediction greatly simplifies inference. We do not need to measure fitness accurately with respect to the broad context that shaped design, a measurement that is usually not possible.

It may not be easy to observe or experimentally create the proper comparison. But without such comparison, there can be no reasonable and broadly applicable way to study the forces of design.

TRADEOFFS

The role of comparison arises in a slightly different way for tradeoffs. For example, it is difficult to say whether a rise in dispersal trades off against a decline in survival. Or whether a pathogen's benefit from increased transmission between hosts trades off against its cost for greater virulent damage to its host. Such associations between traits depend on the underlying mechanism and various confounding factors.[143]

Consider anthrax as an example of the coupling between virulence and transmission. In some habitats, the dominant mode of anthrax transmission is via airborne spores. After pulmonary infection, the pathogen causes few symptoms during the early stages of spread within the host. Only when the bacteria rise to very high density within the host does anthrax express its severe toxins. Those toxins rapidly kill the host. A dead mammalian carcass dries up and releases vast numbers of spores. The severe virulence relates directly to the mode of transmission.

This analysis of anthrax in terms of a virulence-transmission tradeoff does not arise from a testable hypothesis confronted with meaningful data. Instead, it describes observations for one example. Inference is always weak in the analysis of a single case.

To study whether a particular tradeoff dominates design, we must express a comparative prediction.[143] For example, as conditions change to increase a hypothesized mechanistic coupling between virulence and transmission, the observed association between those components of fitness should increase.

The challenge here is to find the correct understanding of the biological mechanisms so that one can accurately predict the coupling between fitness components. For example, we might argue that airborne transmission imposes a stronger virulence-transmission tradeoff than does vector-borne transmission. The data may prove us to be either right or wrong about that comparative prediction.

A similar problem concerns the tradeoff between survival and dispersal. Various microbial mechanisms facilitate dispersal, such as the secretion of molecules that aid movement over surfaces. In some situations, the cost of expressing dispersal-related traits may reduce survival. Comparatively, the more patchy resources are over time and space, the more strongly microbes may be favored to trade local survival for dispersal to new patches.

The tradeoff between survival and dispersal also depends on the cost associated with the particular mechanism of dispersal. For a low-cost mechanism, the association between survival and dispersal may be weak because the marginal loss in survival for an increase in dispersal is small. Thus, wide variation in dispersal rate may be only weakly associated with variation in survival and in overall fitness.

The low association between dispersal and fitness would lead to a weak signal in comparative hypotheses. By contrast, strongly associated marginal changes in dispersal costs and fitness benefits would lead to a strong signal in comparative hypotheses.

In summary, we should consider tradeoffs as traits to be studied directly. As for any trait, we make a comparative prediction. Does an

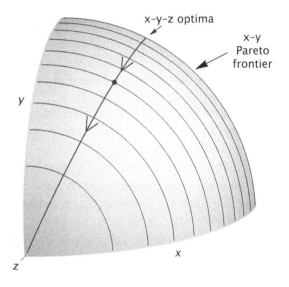

Figure 3.1   Each pairwise tradeoff embeds within a multidimensional tradeoff, confounding the direct study of pairwise tradeoffs.[300] This plot shows tradeoffs between three traits. Suppose, for example, that x is growth rate, y is biomass yield per unit of food intake, and z is toxin production to kill competitors. Assume fitness is $x + by + cz$, with the constraint that $x^2 + y^2 + z^2 = 1$. For fixed values of z, the optimal (x, y) pairs lie along the surface curves perpendicular to the arrow line, often called the Pareto frontier. Comparatively, greater b values predict an increase in y and a decrease in x because of an x versus y tradeoff. The additional z dimension leads to the additional comparative prediction that, as c increases, the optimal (x, y) pairs follow along the surface curve on the unit sphere in which both x and y decrease, as shown in the curve with arrows for $b = 2$. Thus, we may see rate and yield both decrease as toxin production is increasingly favored. If we did not know about toxin production, it would seem as if observed variation were moving orthogonally to the rate versus yield tradeoff, apparently contradicting the existence of that tradeoff. As long as we focus on explicit comparative predictions in terms of changes in b or changes in c, we have a reasonable chance of matching predictions to the forces that shape design.

altered circumstance predict an increase or decrease in the strength of the tradeoff and its importance in shaping design? Comparative predictions about the direction of change provide the only simple, consistent approach to theory and empirical tests (Fig. 3.1).

The essential role of comparison in the study of biological design is not a new idea.[173] Darwin emphasized comparison throughout his work. His focus on comparison was one of his great conceptual innovations. Comparison allowed him to revolutionize the analysis and interpretation of historical processes as forces that shape the observed diversity of biological design.

3.2   Evolutionary Response versus Organismal Response

Comparison focuses on change in trait expression. For example, a microbe may increase siderophore secretion in response to an increase in its genetic relatedness with its neighbors.

Traits change in two different ways. First, increased siderophore secretion may be an evolutionary response to changed conditions. The altered conditions favor genetic change in the population, causing an increase in secretion.

Second, increased secretion may be a phenotypically plastic response of individuals to changed conditions. Individuals sense altered conditions and change their trait expression in response. The response function maps each condition to an expression pattern. Evolutionary forces shape the response function.[82,321,443]

In some simple comparative predictions, we may be able to ignore the distinction between the evolutionary change in trait values and the evolutionary change in trait response functions.

For example, increased relatedness between neighbors will tend to favor greater public goods secretions in both cases. In the first case of fixed trait expression, evolutionary forces will tend to favor genetic change in the population that increases trait expression.

In the second case of phenotypically plastic trait expression, evolutionary forces will tend to favor genetic change in the response function. The favored response function will typically map low relatedness to relatively low expression of public goods secretion and high relatedness to relatively high expression.

In both cases, a change from low to high relatedness predicts increased expression. Evolutionary forces ultimately shape trait expression in the same way. We arrive at the same comparative prediction.

In this simplest description of evolutionary forces and comparative predictions, we do not have to distinguish between evolutionary response and phenotypically plastic organismal response.

The most general approach considers all traits as arising from response functions. Genetically fixed traits sit at one endpoint, such that, for a given genotype, all environments map to the same trait expression. Maximally plastic traits sit at the other endpoint, in which each environment maps to a different expression for the trait.

Must we pay attention to the lability of the response function when formulating comparative hypotheses? The ideal answer is: the more we can do so, the better. The pragmatic answer remains an open problem.

I will often ignore the distinction and focus on comparative predictions for trait values. In some cases, such as explicit consideration of design in the control of plastic trait expression, the response function becomes the focus of analysis.

Future work will need to clarify the limits on simplification. In my view, it is better to start too simply and then add necessary complexity, rather than to start with too much complexity and then subtract to find minimal sufficiency. It is easier to see what is missing in simplicity than it is to see what is not needed in complexity.

3.3   Fundamental Forces and Partial Causes

In physics, gravity is a fundamental force that acts as a partial cause of motion. Gravity by itself rarely provides a complete explanation of motion. In this case, a fundamental force is a component of the various causes of motion.

In biology, I use *fundamental force* to mean a widespread evolutionary process that shapes organismal design. Only the component forces of natural selection can give rise to design. Other evolutionary forces act as constraints with respect to design.

For example, ephemeral resource patches favor dispersal. Stable resource patches favor local survival. These examples illustrate how resource patch demography influences the relative reproductive value of dispersal and survival. Put another way, resource patch demography is a partial cause of design, mediated by the fundamental force of reproductive value.

When considering the design differences between two microbes, several fundamental forces likely play a role. For example, increased genetic correlation between neighbors alters the fundamental force of kin selection that shapes cooperative traits. A cooperative trait might be secretion of exoenzymes for external digestion or secretion of other shareable public goods.

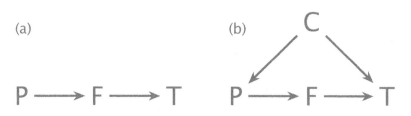

(a)   (b)

Figure 3.2   Partial causes and causal inference. (a) A partial cause in isolation. The parameter, P, influences the trait, T, mediated by the force, F. (b) A covariate, C, causes an association between the parameter and the trait independently of the mediating force. If the confounding effect of C is not resolved, one may obtain an incorrect estimate for the causal effect of P on T.

Increased genetic correlation may influence two opposing forces. On the one hand, increased correlation enhances the shared genetic interests of neighbors, favoring greater expression of cooperative traits. On the other hand, increased correlation raises the competition between related genotypes, favoring lesser expression of cooperative traits.[6,112,333,404,449]

In this case, each force acts in a pathway of partial causation. The first partial cause favors higher trait expression. The second partial cause favors lower trait expression. The net effect may be one way or the other, or no change at all. I develop examples of opposing partial causes in subsequent chapters.

Decomposition of change into partial causes provides significant insight into the forces that shape design, perhaps the best insight that we may achieve.

3.4   Causal Inference

Figure 3.2a shows a pathway of partial causation. The parameter, P, influences the microbial trait, T, mediated by the force, F. Partial causation expresses a comparative prediction. A rise in P predicts the direction of change in T. The predicted direction of change depends on how the pathway of partial causation functions.

Partial causation is always embedded within a larger set of causes. In Fig. 3.2b, the covariate, C, influences both the parameter and the trait. Some of the observed association between the parameter and the trait arises from the common cause, C, rather than the focal pathway, $P \to F \to T$. Thus, it is often misleading simply to measure how observed variation in a parameter associates with observed changes in a trait.

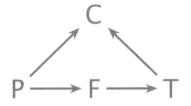

Figure 3.3   An intervening variable, C, is often mistaken for a covariate, leading to incorrect inference about causality. In Elwert & Winship's example, given in Pearl & Mackenzie,[312] we equate the potentially confounding paths, P → C ← T, with three features of movie actors, Beauty → Celebrity ← Talent. Beauty and talent contribute to celebrity. However, beauty and talent are not associated with each other in the general population. Thus, it would be a mistake to treat C as a covariate when analyzing the relation between P and T, because C does not cause an association between P and T.

To isolate the parameter P's partial causative effect on the trait, T, we must remove the causal arrows leading into P, or adjust for those causes. Four approaches isolate the causal effect.[312]

First, controlled randomization of the "treatment" values of P clears potential associations of P with T via indirect confounding causes. In other words, randomization breaks the pathways with arrows going into P. However, it may not be possible to conduct a suitable randomized controlled experiment for the evolutionary response of a trait to changed parameters and evolutionary forces.

Second, one can estimate the causal effect of the parameter separately for each level of each covariate. Conditioning on covariate values removes their potential confounding effect. However, one must successfully identify all of the covariates and then measure them.

The analysis of covariation also requires that one distinguish between a covariate as a common cause, P ← C → T, as in Fig. 3.2b, and an intervening variable that is caused by both the input and the response, P → C ← T, as in Fig. 3.3.

In Fig. 3.3, the variable C does not create an association between P and T. Correcting for C induces a spurious correlation between those variables,[312] as explained in the caption of Fig. 3.3.

Mistaken correction for an intervening variable happens when the direction of causation is ignored. Direction of causality is rarely considered in standard analyses of covariation, leading to many mistaken inferences.[312]

In the third approach to isolating the partial causative effect of P, one directly manipulates the value of P, breaking any incoming causal arrows. The change has to be maintained for a sufficiently long period to achieve a full response. In addition, the intervention becomes a new causal influence into P and so must not itself be correlated with T via a confounding pathway.

The fourth approach, counterfactual analysis, estimates how an imagined manipulation of P would change T. In this case, the observed data without manipulations of P can sometimes provide information about how such a change in P would be expected to alter T. The analysis is *counterfactual* because it runs counter to the factual value of P. Counterfactual analysis has greatly advanced in recent years.[283,312]

Those four approaches describe the methods emphasized by Pearl & Mackenzie.[312] However, in the study of natural history and biological design, one sometimes has to use a weaker observational approach.

If one can aggregate several observed cases of variable P, then any causal paths into P may be sufficiently randomized to reveal the partial causal effect of P on T. Correcting for known correlates, such as common ancestry, strengthens causal insight.

These details support the claim that one can study partial causation in the context of a larger set of causes. The pathways of partial causation are the building blocks of causal understanding.

This book develops hypotheses of partial causation for microbial design. How, in theory, do fundamental forces shape microbial traits? How do environmental and biotic parameters influence those fundamental forces? The broad set of comparative predictions builds the necessary foundation for future progress.

3.5   Structure and Notation of Comparative Predictions

Comparative predictions have the form

$$\Delta \text{ parameter} \Rightarrow \Delta \text{ force} \Rightarrow \Delta \text{ trait}. \tag{3.1}$$

A change in a parameter changes a fundamental force, which changes a trait.

A comparative prediction emphasizes how the direction of change in a parameter predicts the direction of change in a trait. A force mediates the causal pathway of change. For example,

$$P \rightarrow F \dashv T$$

states that an increase in the parameter P causes an increase ($\rightarrow$) in the mediating force F, and an increase in F causes a decrease ($\dashv$) in the trait T. The positive and negative effects multiply through a pathway, so the combination of positive ($\rightarrow$) and negative ($\dashv$) effects yields a net negative cause $P \dashv T$.

Two negative effects, $P \dashv F \dashv T$, yield a net positive cause, $P \rightarrow T$. When a causal effect may be up or down depending on the context, we write $P \dashv\!\!\!\cdot T$.

The example

$$\text{patch lifespan} \rightarrow \text{mutant overgrowth} \dashv \text{secretion}$$

states that the longer resource patches last (Δ parameter), the greater the selective pressure favoring novel mutants that can overgrow the local population (Δ force). Overgrowth mutants may gain their advantage by not secreting external, shareable factors, such as exoenzymes used to digest complex carbohydrates (Δ trait). Reduced secretion of publicly shared factors saves the cost of production.

In comparative predictions, changed parameters act as partial causes of changed traits, mediated by particular forces.

3.6 Recap and Goal

Let's restate the problem. Precise measurements of clonal population growth or other particular components of genetic transmission do not translate directly into the understanding of organismal design. Organismal success has many components that must be combined into an overall fitness measure.

Components of fitness often vary over temporal and spatial scales. One cannot measure all aspects. When evaluating a particular microbe and its design, almost always one will be missing a key component of success.

Focusing on comparison and the fundamental forces often solves the problem. When considering the differences in design between similar microbes, what is the single most important force that explains the observed differences? This question has four aspects.

First, faced with the inability to measure everything that matters, comparison is the only way forward. Comparison focuses measurement on the changes in parameters and forces that cause differences in organismal design.

Second, in the study of design, usually one compares small differences between similar organisms rather than vast changes between different kinds of organisms. Partial analysis of how specific forces alter particular traits makes sense.

Third, among the many forces that may explain differences in design, only a few forces can be the most important ones. It would be ideal to know all of the causal forces. But it is wise to aim first for understanding the most important partial causes in commonly occurring contexts.

Fourth, causal inference provides methods to study pathways of partial causation. This book focuses on theoretical predictions for partial causation rather than on inference. However, it will be important to connect the predictions to empirical study through the methods of causal inference.

I turn now to my main goal, the development of causal hypotheses.

4 Brief Examples

To offer a precise and defendable causal effect estimate, a
well-specified theory is needed to justify assumptions about
underlying causal relationships.

—Morgan & Winship[283]

Comparative predictions arise from fundamental forces. This chapter
presents brief examples.

The first section focuses on metabolism. The fundamental forces
shape the tradeoffs between growth rate, reproductive yield, and other
components of fitness. Testable comparative predictions follow, revealing the forces that shape metabolic design.

The second section discusses how to test comparative predictions.
A comparative prediction describes the expected direction of change
in a trait in response to a change in conditions. Support follows when
the overall tendency of observed change in the trait is in the predicted
direction.

The third section describes microbial "cancers." When an isolated
genetic clone grows in a long-lasting resource patch, mutants arise that
increase growth rate. The mutants overgrow their clonemates because
of their short-term competitive advantage, like a tumor overgrows its
surrounding tissue. However, the mutants often disappear over the long
term because of reduced overall fitness.

The fourth section analyzes heterogeneity in the production of secreted molecules. The secreted molecules include iron-scavenging siderophores, glycan-digesting exoenzymes, and quorum-sensing signals.
Once a molecule is secreted, the benefits may be shared publicly by
neighboring cells. Relatively vigorous cells may produce more of the
publicly shared goods because vigorous cells suffer lower marginal costs
of production when compared with less vigorous cells.

The fifth section evaluates the stages in the lifespan of a resource patch. At early stages, a secreted molecule provides cooperative benefit to the current and several future generations. At later stages in the patch's lifespan, a secreted molecule provides less future benefit. Cooperative traits may decline with the age of a patch as the marginal gains decay. In general, demography powerfully shapes the design of traits.

Subsequent chapters develop many comparative predictions. Those comparative predictions relate fundamental forces to microbial design, providing the basis for future empirical and theoretical studies.

4.1   Metabolism and Growth Rate

Recent studies focus on the design of microbial metabolism.[364] The question arises: Designed to achieve what particular goal?

One possibility is growth rate. By growing faster, a microbe outcompetes its neighbors. However, microbes often do not maximize growth rate.[27] Various mutants grow faster. Natural selection does not favor the full potential for growth maximization.

If microbes are not evolving to maximize growth rate, then what is the design target?[364]

EXAMPLE TRADEOFF: RATE VERSUS YIELD

Perhaps growth rate trades off against yield.[317,444] Roughly speaking, yield is the conversion efficiency of available resources into reproductive output or biomass.

The idea is that growing faster uses resources to increase speed, reducing the resources available to produce offspring. There is a tradeoff between rate and yield.

For example, a microbe can make more of a cell surface transporter to increase its rate of nutrient acquisition.[453] The extra resources devoted to nutrient transport reduce the net yield at which the nutrient is changed into biomass.

Figure 4.1 shows comparative predictions for the rate-yield tradeoff. Those predictions arise from the analysis of specific mathematical models.[130] I emphasize two predictions.

Mixing ⊣ relatedness ⊣ rate. Increased genetic mixing of colonists in a resource patch lowers the relatedness (genetic similarity) between neigh-

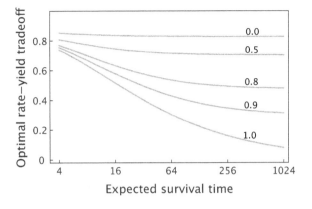

Figure 4.1   Rate versus yield tradeoff in microbial metabolism. Increasing height corresponds to greater growth rate and lower efficiency yield of reproductive biomass per unit resource input. The text explains the roles of genetic related-ness (numbers above curves) and expected patch survival time (x-axis). Patches begin with a fixed amount of resource. There is no further resource influx. Time is expressed in dimensionless units. From Figure 1 in Frank.[130]

boring cells within a patch. Lower relatedness favors faster growth rate to outcompete neighbors and gain genetic representation in future generations, a rise in fitness. By contrast, when relatedness is high between neighbors, then outcompeting genetically similar neighbors by faster growth provides little gain in terms of future genetic representation in the population.[130,317]

In Figure 4.1, the number above each curve is the genetic relatedness between neighbors within a patch. As relatedness declines, the curves increase in height. The height represents the fraction of resources devoted to faster growth rate. More rapid growth lowers yield.

The large increase in growth rate as genetic relatedness declines shows the dominant influence of population genetic structure on metabolic design. Genetic relatedness associates with the fundamental force of kin or similarity selection.

Patch lifespan → marginal yield ⊣ rate. Increased lifespan of resource patches enhances the marginal benefit of yield efficiency, which decreases growth rate. In a long-lived patch with limited resources, all resources may be consumed. The longer patches last, the greater the marginal fitness gain from increased yield efficiency.

By contrast, a short-lived patch favors fast growth because the patch will disappear before all of the available resources are used up. With excess resources during the patch lifespan, metabolic efficiency provides relatively little marginal gain in fitness.

Figure 4.1 illustrates the effect of patch lifespan on metabolic design. Each patch begins with a fixed amount of resource. As average patch lifespan increases, the probability of resource depletion rises. Resource depletion favors greater yield and reduced growth rate.

The figure shows the declining growth rate with increasing lifespan. Variable patch lifespan associates with the changing fundamental force of demography.

These rate versus yield predictions demonstrate the power of key environmental parameters to shape the design of microbial metabolism. Changes in those parameters alter the fundamental forces of similarity selection and demography. Simple comparative predictions follow. Those comparative predictions provide testable hypotheses.

PROBLEM: OTHER TRADEOFFS

However, a model focused solely on rate versus yield can mislead. For example, shorter patch lifespan may greatly increase the importance of dispersal. If so, declining patch lifespan may associate with greater dispersal, lower growth rate, and reduced yield.

We could analyze dispersal as another tradeoff with rate and yield. What about tradeoffs with toxin production, defense against attack, and survival by production of quiescent and long-lived spores? We could include those. But a model of everything is a model of nothing.

SOLUTION 1: COMPARATIVE PREDICTIONS FOR TRAITS

We will do better by focusing on the trait of interest rather than all of the possible tradeoffs. How does changing genetic relatedness alter growth rate? How does patch lifespan alter growth rate?

We cannot predict the expected changes exactly without knowing all of the tradeoffs. But we can often predict the direction of change. With that in mind, let's look again at the predictions about growth rate from the earlier subsection. When necessary, we modify those predictions to avoid stating tradeoffs explicitly.

Mixing ⊣ relatedness ⊣ rate. Genetic mixing reduces relatedness, which predicts an increase in growth rate. Here, we do not invoke a tradeoff. The prediction focuses entirely on how competition between increasingly different types favors faster growth.

In this case, we make an implicit assumption about tradeoffs in relation to more genetic mixing. The gains in fitness from increased growth rate are greater than the associated losses in other components of fitness.

This assumption will probably be true in many cases. Competition between types directly and powerfully affects fitness. Other components of success tend to be less direct.

We may sometimes be wrong. But we aim only to be right about the direction of change in a trait significantly more often than we are wrong.

Meanwhile, we can study many different models of tradeoffs. How often does the theory support our simplifying assumptions? In empirical tests, how often do observations support the directional prediction? We revisit these issues in Chapter 10.

Patch lifespan ⊣ marginal rate → rate. When time is short, it pays to grow fast. In other words, as patch lifespan decreases, the marginal gain in growth rate rises. Here, each ephemeral patch begins with a fixed amount of resource, which may be consumed over the patch lifetime.

The second part of the comparative prediction says that the more strongly the marginal gain for growth rate changes, the more strongly the growth rate itself changes in the same direction.

Without a specific tradeoff, the argument is that shorter patch lifespan typically favors more rapid growth. That argument concerns a partial pathway of causation, ignoring other possible causes.

When considering all potential causes, some tradeoffs may lead to a different outcome. For example, shorter patch lifespan may strongly favor increased dispersal at the expense of reduced growth rate. The net effect of shortened patch lifespan on growth rate depends on the balance of alternative forces through different pathways of causation.

Over all tradeoffs, the total causal effect of shortened patch lifespan on growth rate will often depend on the details of several causal pathways.

Despite potential complications, reduced lifespan of fixed resource patches does favor faster growth when considered as a partial pathway of causation. We should not lose sight of the likely possibility that, over all reasonable conditions, the simple qualitative prediction about the direction of change makes theoretical sense and may be supported by empirical test (Section 4.2).

SOLUTION 2: COMPARATIVE PREDICTIONS FOR TRADEOFFS

We often do not know which tradeoffs dominate. To solve that problem, the previous subsection aggregated over various potential tradeoffs. Over all likely tradeoffs, what is the general tendency for the direction of change in a trait? Such predictions do not depend on the details of particular tradeoffs.

However, we often wish to understand the tradeoffs that shape design. How can we study those tradeoffs?

The solution always comes back to formulating a comparative prediction. How does change in some parameter alter the tradeoff?

Consider an example prediction. As sugar becomes more limiting, the tradeoff between growth rate and reproductive yield strengthens.

This example predicts the intensity of the tradeoff. Low sugar availability, which imposes free energy limitation, enhances the free energy tradeoff between rate and yield. By contrast, high sugar availability provides excess free energy, which reduces the relative importance of the free energy tradeoff between rate and yield. An overall negative rate-yield association may still occur under excess sugar, but the negative association may become weaker and more variable.

The actual change in tradeoff intensity depends on the underlying mechanisms that couple traits and on the abundances of the various resources that limit growth. However, aggregating over different underlying mechanisms and levels of other resources, we may expect a general tendency for increased sugar availability to make the negative rate-yield association weaker and more variable.

Put another way, with excess sugar, some other nutrient probably becomes limiting, which strengthens an alternative tradeoff that may limit growth rate or yield efficiency.

The more specific we can make comparative predictions in relation to the underlying mechanisms and the levels of other potentially limiting resources, the more we can claim that support for those predictions reveals the forces that shape design.

In summary, comparative predictions provide a direct way to study the tradeoffs that shape design. In theory, how do changed conditions alter the relative strength of particular tradeoffs? For such comparative predictions, do we observe support by empirical test?

4.2   Support by Empirical Test

I used the phrase *support by empirical test* with regard to comparative predictions about particular traits or tradeoffs. The phrase means that, over many different situations, the overall tendency of observed change will be in the predicted direction. Simple comparative predictions and empirical tests provide the only widely applicable way to study the fundamental forces of design.

I will not say how such empirical tests should be done. That problem is not an easy one. But I am confident that, once the conceptual challenge is clearly understood by the broad community of scientists, progress will follow.

In general, two factors set the rate of progress on big problems. Clarity of conceptual framing focuses attention. Innovation in empirical study and measurement technology opens the way to real progress.

Neither factor always leads. Novel measurement technologies provide new kinds of data, which set new conceptual puzzles. Conceptual progress, in turn, demands new kinds of measurement. My claim is that, in the study of microbial design, conceptual issues currently limit progress.

As the conceptual approach builds, the demand will naturally shift back to the empirical side. The scientific community is fantastically good at solving empirical challenges once compelling conceptual challenges have been presented.

4.3   Patch Lifespan and Microbial Cancer

Natural selection often tunes microbial metabolism to growth rates below the maximum that could be achieved.[27] Section 4.1 showed that trading reduced growth rate for increased yield may increase fitness. In that analysis, long-term fitness includes a full demographic cycle. A cycle begins with growth in temporary resource patches and completes with dispersal to colonize newly arising resource patches.

If a resource patch lasts long enough, mutants with higher growth rates will likely arise within that patch. For example, a gene duplication of a transporter can increase the uptake rate of a limiting resource and enhance growth rate.[204,470] A mutant with faster reproductive rate overgrows neighbors and dominates the local patch.

In a patch initially colonized by a single genetic clone, an overgrowth mutant is like a microbial cancer.[135] The population begins as a genetically uniform multicellular soma. The fast-growing mutant then spreads in the "body" like a cancerous tumor.

TIMESCALE

The fast-growing mutant gains fitness in the short term within its patch.[236] But it may lose fitness over the long term because fast growth often trades off against other components of fitness, such as survival or yield.[84] Low yield reduces the total number of progeny that can potentially disperse to colonize new patches. Over a complete demographic cycle, a fast-growing mutant may have lower total fitness than a genotype with slower growth and higher yield.

This conflict between fitness components acting on different timescales can influence the design of metabolism.[129,130,133,135] For example, relatively short patch lifespans typically do not allow sufficient time for new mutants to arise and overgrow the population. In that case, the long timescale over the full demographic cycle dominates. The analysis in Section 4.1 focused on that long timescale.

By contrast, relatively long patch lifespans with continual resource influx emphasize the mutant overgrowth process within patches. The nonmutant descendants of the initial colony progenitors inevitably face strong selective pressure from their fast-growing mutant siblings.

The within-patch selective pressure from mutants favors the initial progenitors to evolve faster growth and consequently lower yield so that their nonmutant descendants can better compete with mutant siblings.[129,130,133,135]

EXAMPLE: RATE VERSUS YIELD

I use yield as an example of an alternative fitness component that trades off with growth rate. Similar ideas about microbial cancers arise with tradeoffs between growth rate and other fitness components, such as dispersal. Yield provides a convenient example because of prior mathematical analyses and simple underlying logic.

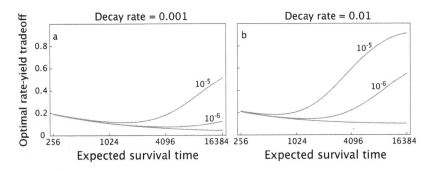

Figure 4.2   Rate versus yield tradeoff as influenced by cancer-like overgrowth mutants. Curves show the predicted rate-yield combination expressed by the initial progenitors that colonize patches. Within a patch over its lifetime, the combination may evolve toward higher rate and lower yield. The axes are the same as in Fig. 4.1, with average patch lifespan expressed as expected patch survival time. In this case, patches have continual resource influx, allowing continual reproduction and cellular death throughout the patch lifespan. Numbers above the curves show mutation rates, with a mutation rate of zero for the lowest curve in each panel. Panels (a) and (b) differ in decay rates, which correspond to both the cellular death rate and the decay rate of the membrane transporters that mediate the growth versus yield tradeoff. Units for all quantities expressed nondimensionally. Full details in Figure 2 of Frank.[130]

Figure 4.2 illustrates the role of changing patch lifespan on the rate versus yield tradeoff. The bottom curve in each panel shows the rate-yield combination favored by selection when there is a single clonal colonist and no within-patch mutation, corresponding to a genetic correlation within patches of 1.0 in Fig. 4.1.

The upper curves in each panel of Fig. 4.2 show increasing mutation rates. Higher mutation rates and longer patch lifespans favor faster growth and lower yield among the initial progenitor genotypes that colonize patches.

Above each panel, the decay rate parameter describes both the death rate of cells and the decay rate of transporters. Faster cellular decay associates with more rapid cellular turnover, which enhances the rate at which overgrowth mutants spread. Faster transporter decay favors more investment in the production of new growth-related transporters, reducing yield and increasing the relative dominance of rate over yield.

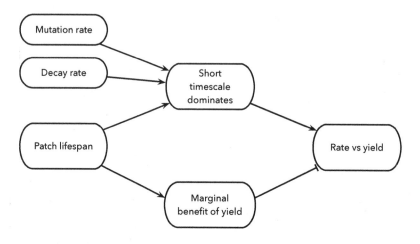

Figure 4.3 Causal pathways by which mutant overgrowth influences the rate versus yield tradeoff in metabolism. These pathways derive from Fig. 4.2, based on the model in Frank.[130]

ALTERNATIVE CAUSAL PATHWAYS

Figure 4.3 shows the various causal pathways that influence the rate-yield tradeoff in metabolism. In the upper pathway, increases in mutation rate, decay rate, and patch lifespan all increase the spread of overgrowth mutations in the local population. The figure describes the increase in mutational overgrowth in terms of the enhanced weighting of the short-term forces favoring growth rate within the patch relative to the long-term forces favoring yield, which dominate the competition between patches.

We can describe the increase in mutational overgrowth in terms of several alternative forces. For example, more mutations reduce genetic similarity because the rise and spread of mutations increase genetic heterogeneity. The initially pure clone becomes a genetic mixture.

Alternatively, more mutations enhance the weighting of short-term forces within the patch relative to long-term forces between patches by increasing the within-patch genetic variation.

Or, the greater competition that arises within patches via fast-growing mutants weights the reproductive value (RV) component of relative growth rate more strongly than the reproductive value component of dispersal to colonize new patches.

ALTERNATIVE COMPARATIVE PREDICTIONS

The different ways of specifying cause lead to several equivalent comparative predictions, each emphasizing a different description of the fundamental driving force of change

> patch lifespan → short-timescale weight → rate
>
> patch lifespan ⊣ relatedness ⊣ rate
>
> patch lifespan → within-patch variation → rate
>
> patch lifespan ⊣ RV dispersal weight ⊣ rate.

Each prediction describes the same partial causal relation between parameter and trait, patch lifespan → rate. In each case, greater patch lifespan increases rate by weighting more heavily the upper pathway of partial causation in Fig. 4.3.

The partial causal relation, patch lifespan → rate, can be described by a variety of fundamental forces that mediate the causal relation. The fundamental force is the middle expression in each of the listed predictions. The fact that alternative descriptions can be used for the same prediction often leads to needless controversy and confusion.

It is best to keep in mind all of the alternative ways for describing forces. Each expression differs slightly in emphasis and potential insight. The biology of a particular problem often makes clear which expression is most helpful.

BALANCE OF OPPOSING FORCES

In Fig. 4.3, increasing patch lifespan also tends to reduce rate via the lower partial pathway of causation, patch lifespan ⊣ rate. The net effect of a change in patch lifespan through those opposing pathways of partial causation depends on the relative strength of the two pathways. Figure 4.2 shows the resolution of opposing forces for particular assumptions.

An ideal empirical analysis studies each partial pathway of causation in isolation. Separation can sometimes be achieved experimentally by controlling the alternative pathway. Separation can sometimes be achieved statistically by developing causal inference procedures for data analysis that correct for the alternative pathway.

This section emphasized the opposition of forces acting over short and long timescales. In nature, those opposing forces likely create diversity in metabolic traits over time and space. Understanding that metabolic diversity remains an open challenge for both theoretical and empirical analysis.

4.4   Heterogeneity in Public Goods

Microbes often secrete molecules. Siderophores scavenge iron. Exoenzymes break down large glycan molecules. Quorum sensing signals provide information to other cells about population density.

Secretion imposes a cost of production. The secreted molecule, once released, often benefits all neighboring cells. In other words, the secretor bears the cost, and the neighborhood gains the benefit of the publicly shared good.

Many studies of microbial public goods have been published. However, those studies rarely evaluate cellular heterogeneity in the capacity to secrete public goods.[216,388,441]

Consider vigorous cells. Producing an additional secreted molecule will cause little harm to other processes. The particular reason for greater vigor or access to resources does not matter.

Struggling cells, with limited energy or resources, pay a relatively larger marginal cost for each additional secreted molecule. Two predictions follow.[127]

Vigor ⊣ marginal costs ⊣ secretion. At the individual cellular level, increasing vigor reduces the marginal costs of production, which favors greater secretion of public goods.

Het vigor → het costs → het secretion. At the population level, increasing heterogeneity (het) in vigor increases heterogeneity in the marginal costs of production, which favors greater heterogeneity in the secretion of public goods.

These abstract predictions lack biological detail. However, the logic should apply to many different secretion traits. Broadly applicable predictions reveal principles for understanding microbial design.

4.5   Stage-Dependent Growth

Some processes may limit patch lifespan. Resources may be ephemeral. Pathogens may be cleared from a host.

A patch's reproductive value rises with the number of dispersers produced over the patch lifespan. A cooperative trait, such as secretion of a public good, can increase population size and total dispersal.

The benefit for secreting public goods depends on the stage in the patch's lifespan. Greater secretion early in the lifetime of the patch may provide population growth benefits that carry forward over several generations. For example, if the decay rate of a secreted exoenzyme is slower than the cellular generation time, then the enzyme continues to function over several generations.

Secretion near the end of the patch lifespan provides relatively limited benefits. Stage-dependent decline in marginal benefits leads to the comparative prediction[128]

$$\text{stage} \dashv \text{marginal benefits} \rightarrow \text{public goods.}$$

Increasing stage reduces the benefits of secreting public goods. Lower benefits decrease secretion. Overall, the predicted secretion rate declines over the course of a patch's lifespan. The stage in a patch's lifespan is similar to age, emphasizing the broad role of demography in shaping the design of microbial traits.

Many factors potentially alter the costs and benefits of traits at different stages in a resource patch's lifespan. Changes in those age-specific factors predict changes in trait expression over a patch's lifetime. Demography is a fundamental force that alters traits.

4.6   Summary

This chapter sketched the logic and structure of comparative predictions. Later chapters develop predictions for particular microbial traits. Those predictions concisely express testable hypotheses about the causal forces of microbial design.

The predictions also provide the basis for future theoretical work. That future theory must develop more realistic analyses for the interaction of multiple forces. Teasing apart the multiple evolutionary processes that act simultaneously is a major challenge in the study of design.

5 Theory: Forces

There is no falsification before the emergence of a better theory.

—Imre Lakatos[220]

Comparative predictions arise from the fundamental forces of design. This part of the book introduces the theory toolbox. I emphasize concepts and demonstrate applications.

Later chapters present many comparative predictions. One can test those predictions without studying the theory. However, a strong theoretical foundation enhances application to challenging problems and opens new predictions.

This chapter focuses on conflict, cooperation, and life history components of fitness. The theory depends on the temporal and spatial scales of success and on variability in performance. The next chapter focuses on the nature, origin, and modification of traits. The final theory chapter introduces design principles for the regulation and control of traits.

The first section of this chapter presents the tragedy of the commons. The forces of design often favor individuals to outcompete their neighbors. Resources devoted to competition reduce the resources available for reproduction. Inefficient use of common resources for reproduction degrades the success of all individuals. This tragedy of the commons powerfully shapes the design of many microbial traits.

The second section demonstrates that similarity between neighbors reduces the tendency to compete and increases cooperative efficiency. Change in neighbors' similarity leads to strong comparative predictions about microbial traits. Similarity sometimes arises by kinship, which correlates genotypes between neighbors. Other processes also influence similarity, modulating a common force that shapes microbial design.

The third section measures the marginal gains and losses in trade-offs. For example, how does natural selection alter traits to balance the competitive gains against neighbors versus the cooperative efficiency of resource use? The balance typically occurs when the marginal gain for slightly better competitive success equals the marginal loss for slightly worse efficiency. If the marginal gain in competition were greater than the marginal loss in efficiency, then selection would alter traits to enhance competition. Traits change until marginal gains and losses balance.

The fourth section shows that repression of competition may enhance efficiency. If a trait prevents neighbors from competing, then individuals can increase their success only by raising the group's shared efficiency. Repression of competition provides an alternative to similarity for reducing competitiveness and increasing efficiency.

The fifth section considers the production of public goods. A public good is something produced by an individual that benefits all neighbors. The producing individual bears the cost, whereas all neighbors share the benefit. A vigorous individual pays a smaller marginal cost of production because a unit of production takes up a smaller fraction of a vigorous individual's total resources. The population splits into vigorous producers and weak nonproducers.

The sixth section weighs different components of fitness on the common scale of reproductive value. A fitness component's reproductive value measures its genetic contribution to the future population. For example, we may consider the tradeoff between faster reproduction in a rare habitat versus slower reproduction in a common habitat. The benefit in the rare habitat must be weighted by the low contribution of that habitat to the future population. Similarly, the cost in the common habitat must be weighted by its high future contribution.

Demographic analysis provides the exact weighting of different fitness components by their projected future contribution. Fitness components include fecundity, survival, dispersal, and success in different habitats. Changes in demographic parameters alter the relative reproductive values of fitness components in tradeoffs. Altered reproductive values cause changes in key traits, leading to strong comparative predictions.

The seventh section evaluates traits expressed at various stages in a colony life cycle. Fitness at each stage depends on survival to that stage multiplied by the number of successful dispersers produced at that stage. Traits that increase survival at an early stage in the colony life cycle have high reproductive value because they enhance the probability of survival to all future stages. By contrast, traits that increase survival only at late stages in the colony life cycle have low reproductive value.

The exact reproductive value weighting of survival and fecundity components at each stage can be calculated by demographic analysis. Changes in reproductive value weightings across stages predict changes in stage-specific trait expression.

The eighth section reviews the three key measures of value. Similarity enhances the value of cooperative traits by increasing the shared interests of neighbors. Marginal values compare how changed trait values alter different components of success. Reproductive values weight fitness components by their relative contribution to the future population.

The final two sections raise additional forces. The spatial and temporal scaling of competition alters the fitness valuation metrics. Variability in performance alters the fitness value associated with a trait. Modulated fitness values change the design of traits.

5.1   Tragedy of the Commons

> In a single battle the Peloponnesians and their allies may be able to defy all Hellas, but they are incapacitated from carrying on a war. ... Slow in assembling, they devote a very small fraction of the time to the consideration of any public object, most of it to the prosecution of their own objects ... and so, by the same notion being entertained by all separately, the common cause imperceptibly decays.
>
> —Thucydides[410]

Cells within a clone share genes. Outcompeting clonal neighbors provides no benefit. The gained genetic transmission is offset by lost transmission of the same genes. With no chance to gain by competition, selection favors clonal traits that use common resources most efficiently.

Genetic diversity breaks common interest. Genotypes gain in the short term by outcompeting neighbors. Better competitors take more of the common resources or contribute less to the common good.

Degrading the commons reduces long-term efficiency. The better competitors initially increase but ultimately do worse over the full demographic cycle. Changed conditions that increase short-term competition between genotypes also lower long-term efficiency.

Thucydides perfectly expressed the conflict between self-interest and group efficiency in this section's epigraph. Hardin[172] named this conflict *the tragedy of the commons.* Frank[115,117] showed that the tragedy of the commons powerfully shapes the design of microbial traits.

5.2   Similarity Selection and Kin Selection

In the tragedy, dissimilar neighbors gain by competing. Increased similarity favors greater cooperation. More cooperation reduces wasteful competition and enhances success for all group members.

A simple tragedy model illustrates the fundamental force of similarity.[114,115,117,122,128] The model begins by writing the expected fitness of an individual, w, in terms of the individual's competitiveness, y, and the average competitiveness of its neighbors, z, as

$$w = \frac{y}{z}(1 - z). \tag{5.1}$$

The relative success of an individual is y/z, the individual's competitiveness relative to its neighbors.

Greater investment in competitiveness reduces efficiency. For example, if individuals invest more in cell surface transporters to extract resources from the commons, the cost of extra transporters takes away from resources that might otherwise have gone directly into reproduction.

Let y and z vary between 0 and 1. Values denote the fraction of maximal competitiveness. Less competitive individuals use resources for reproduction more cooperatively and efficiently. Thus, the group efficiency increases with $1 - z$.

Individual fitness in eqn 5.1 is proportional to the group success, $1 - z$, multiplied by the relative success of an individual within the group, y/z. In other words, efficiency in use of the commons sets the total success of the group. Competitiveness of individuals against their neighbors sets their share of the total.

Here, *neighbors* means those individuals in the neighborhood, which includes the focal individual. For example, if the neighborhood has two individuals, including the focal individual, then one-half of the average value in the neighborhood comes from the focal individual.

ANALYSIS OF SIMILARITY

What level of competitiveness maximizes individual gain? Under simple assumptions,[115,122,405] we find the maximum by setting to zero the derivative of w with respect to individual competitiveness, y. Using the standard chain rule of differentiation yields

$$\frac{dw}{dy} = \frac{\partial w}{\partial y} + \frac{\partial w}{\partial z}\frac{dz}{dy} = 0. \tag{5.2}$$

Notation aids interpretation. Let

$$-C_m = \frac{\partial w}{\partial y} \qquad B_m = \frac{\partial w}{\partial z} \qquad r = \frac{dz}{dy},$$

so that

$$\frac{dw}{dy} = -C_m + rB_m = 0. \tag{5.3}$$

The term $-C_m$ is the direct effect of the focal individual's phenotype on its own fitness. Traditionally, in these models, one studies altruistic traits that reduce the focal individual's fitness and provide a benefit to its neighbors. Examples include secretion of siderophores and secretion of exoenzymes. The marginal cost for the altruistic trait to the focal individual is C_m, in which a positive cost reduces fitness.

The term B_m is the direct effect of the average group phenotype on the focal individual's fitness. In a model of altruism, B_m is the marginal benefit to the focal individual for an increase in the average altruistic trait expression of neighbors. For example, the focal individual gains as its neighbors increase their secretion of public goods, such as siderophores and exoenzymes.

The value of r is the slope of the group phenotype on the focal individual's phenotype. That slope measures the similarity between the focal individual and its group.

If we let $c = C_m$ and $b = B_m$, an increase in an individual's trait enhances its fitness when $dw/dy > 0$, which occurs when $rb - c > 0$. That condition has the same form as Hamilton's rule for the spread of an altruistic trait.[166-168] Technically, Hamilton's rule differs from this expression, although the interpretation is similar (Section 15.6).[120-122]

When trait values are at maximum fitness, individuals cannot do better by changing their phenotype. For simple assumptions, that means that everyone must have the same phenotype, $y = z = z^*$, and any deviants do worse.[122] Therefore, we can find the maximum by solving eqn 5.3 at the point $y = z = z^*$.

Assuming that all individuals have the same phenotype is not realistic. But the goal of this model is not to match reality. Instead, we clarify how various forces act to shape the design of phenotypes. If we can identify the broad characteristics of those forces, then we can make testable comparative predictions.

Eqn 5.3 applies when we can write individual fitness, w, in terms of individual phenotype and group average phenotype, y and z. The equation suggests how various forces shape the design of traits. In particular, the force of similarity expressed by r interacts with the forces of marginal costs and benefits, C_m and B_m.

APPLICATION TO THE TRAGEDY

The tragedy of the commons illustrates the role of similarity in the balance of forces. Applying the methods in the previous subsection to eqn 5.1 yields

$$-C_m = \frac{1 - z^*}{z^*} \qquad rB_m = -\frac{r}{z^*}.$$

From eqn 5.3 we obtain[114,115,117]

$$z^* = 1 - r. \tag{5.4}$$

The competitiveness, z^*, rises toward its maximum value as r becomes small and the similarity of an individual to its group declines.

When all individuals express the maximal fitness trait value, z^*, their fitness is $1 - z^* = r$. That result follows from the fitness expression in eqn 5.1 evaluated at the maximum, $y = z = z^*$.

Thus we have the tragedy. Lower similarity, r, favors greater competitiveness, z^*. Everyone's fitness decreases because fitness is $1 - z^*$.

Greater similarity rescues the tragedy: as r increases, z^* declines. Everyone's competitiveness declines and their fitness rises.

We could have a more complicated functional relation between competitiveness and fitness. The tragedy remains whenever greater competitiveness enhances individual success relative to neighbors and greater competitiveness degrades the efficiency of the group.

In summary, similarity modulates the design of competitive traits. Greater similarity between neighbors alleviates the tragedy, reducing competitiveness and increasing individual and group success.

APPLICATION TO PUBLIC GOODS

A public good is something produced by an individual that benefits all neighbors. The producer pays the cost of production. Everyone gains the benefit. An individual that reduces its production lowers its own cost but still shares in the benefits of public goods produced by others. Cheating nonproducers raise their competitiveness against their neighbors.

We can match the public goods problem to the fitness expression in eqn 5.1. Lower individual production raises competitiveness against neighbors. Thus, we can think of z as reduced production and $1 - z$ as the average level of public goods production. Then the favored value of public goods production from eqn 5.4 is

$$1 - z^* = r.$$

Once again, greater similarity, r, favors more cooperation. In this case, cooperation means the level of public goods production, $1 - z^*$.

Other mechanistic assumptions lead to different fitness expressions. For example, suppose $1 - y$ is the public goods productivity of a randomly chosen individual, and group success is proportional to the average public goods productivity, $1 - z$. In this case, assume that an individual's reproductive vigor is equal to one minus its public goods production, $1 - (1 - y) = y$.

Here, vigor directly affects an individual's reproduction rather than affecting its ability to compete with neighbors. Fitness is an individual's intrinsic vigor, y, multiplied by the neighborhood's quality determined by its public goods productivity, $1 - z$, such that

$$w = y(1 - z).$$

Following the standard procedure, the favored level of public goods production is

$$1 - z^* = \frac{r}{1 + r}. \tag{5.5}$$

We have the same qualitative comparative result. Rising similarity, r, predicts increased public goods production, $1 - z^*$.

PHENOTYPIC SIMILARITY

The similarity in the prior subsections depends only on phenotype. The individuals could be members of the same species or members of different species.

For example, the habitat may be divided into many small resource patches. In each patch, two individuals may compete for a common resource. Similarity describes, on average, how closely matched the level of competitiveness is between patchmates. Similarity is high when strong competitors tend to match with strong competitors and weak competitors with weak competitors. The weak competitors can be thought of as strong cooperators.

The cause of matching does not matter. It may be that, for members of different species, strong cooperators use similar cues to find resources. Or, for members of the same species, similar phenotypes may be associated spatially because they share common genotypes.

Whatever the cause of phenotypic similarity, the association within patches influences success in reproduction. In the tragedy model, the more similar the trait values, the more individuals are favored to reduce their competitiveness and share in enhanced efficiency.

GENOTYPIC SIMILARITY TO NEIGHBORS' PHENOTYPE

Success in reproduction only influences evolutionary pattern when the associated traits transmit to future generations. If the descendants do not inherit the successful traits, then no evolutionary response occurs.

In the prior phenotypic analysis, we focused on an individual with trait y in a group with average trait value, z. Suppose our focal individual has a combination of heritable factors that, on average, causes their bearers to express trait values, g, such that[121,122]

$$y = g + \epsilon. \tag{5.6}$$

The trait value depends on the genetic value, g, plus an environmental or error term, ϵ. The average of ϵ is zero. The average trait value is the average genetic value, $\bar{y} = \bar{g}$. Here, *genetic* means transmissible factors that influence trait expression.

Changes in g determine the evolution of average trait values. We can rewrite the general expression for trait evolution in eqn 5.2 as

$$\frac{dw}{dg} = \frac{\partial w}{\partial y}\frac{dy}{dg} + \frac{\partial w}{\partial z}\frac{dz}{dg} = 0. \tag{5.7}$$

From eqn 5.6, we have $dy/dg = 1$. Let $r = dz/dg$, the slope of average group phenotype on the transmitted genetic value of the focal individual.

Then we recover all of the results above but with r now describing the transmitted component of phenotype.[122]

Causes of Similarity

In many cases, similarity between the focal individual and its neighbors arises because of genetic similarity. For example, the group may share a recent common ancestor, forming a kinship group. Or the genetic similarity may arise because similar genotypes tend to aggregate spatially, independently of common descent.

The method described here does not depend on the cause of similarity, r, between the neighbors' average phenotype and the focal individual's genetic value. Restrictive assumptions about similarity arise in other methods, such as Hamilton's inclusive fitness, strictly defined notions of kinship, or group selection.[122,136]

Sometimes the more restrictive assumptions and methods provide special insight. However, in most cases, one gains little practical value by the special assumptions of the other approaches. Here, I will use the simpler and typically more general analysis, leading to expressions such as eqn 5.7.

Mutation Degrades Similarity

A single cell may colonize a resource patch. As the clone expands, the genetically identical cells have perfect similarity, $r = 1$. Inevitably, mutations arise within the clone, degrading similarity. As mutations increase in frequency and r declines, the cohesive force of similarity breaks down. Selection favors greater competition within the group, reducing cooperation and efficiency.

Common descent and kinship set a tendency for similarity and cooperation. But the true causal force depends only on current similarity. Past history does not matter for selection. Only current phenotype and future genetic transmission matter. "The Pedigree of Honey/Does not concern the Bee."[85]

If mutation and selection enhance phenotypic similarity, then r rises even though nucleotide divergence increases. Once again, the causal force depends on the phenotypic similarity of neighbors relative to the actor's genetically transmissible trait value, measured by r, rather than on history, kinship, or nucleotide similarity.[122]

CHOICE OF PHRASE: SIMILARITY SELECTION OR KIN SELECTION

Similarity selection provides the most accurate phrase. However, similarity often arises by common descent and kinship. *Kin selection* is widely used. Few people recognize *similarity selection.*

Anyone who understands the basic principles should immediately recognize the historical broadening of concepts that derived from the original usage of *kin selection.*[136] However, the literature strongly suggests that *kin selection* ignites wasteful controversy and misunderstanding.

Social selection is sometimes used[280,352] for the same concepts as *similarity selection,* emphasizing that similar social partners do not have to be kin.[113,120,122] However, *social selection* is also used in other contexts,[255,442] and the phrase does not emphasize the essential factor of similarity.

In this book, I will use both *similarity selection* and *kin selection,* which I regard as interchangeable phrases.

COOPERATION BETWEEN SPECIES

For the r coefficient in eqn 5.7, only the average phenotypic value of neighbors matters. Those neighbors, comprising z, may be from different species.[113] The "relatedness" coefficient is the slope of the average phenotypic value of neighbors on the focal individual's genetic value.

The fact that correlated trait values between different species can drive cooperation between those species raises interesting questions in theory and application.[73,87,110,113,116,119,170,171,460]

Are correlated traits between species necessary for cooperative evolution between species? That depends on how one defines cooperation between species. Suppose, for example, that species A gains from something produced by species B. Then, if an individual of species A, at a cost to itself, provides something that enhances the growth of species B, the additional beneficial product made by species B returns a benefit to members of species A.

In this case, species B acts passively as a part of the environment, and we do not consider this as an evolutionary problem of cooperation between species. But, in terms of ecological process, we may sometimes wish to view this process as an aspect of mutual benefit between species because A is enhancing B's growth, and B is enhancing A's growth.

Evolutionarily, the problem concerns which individuals in species A receive the return benefit. If, initially, a particular individual of species A performs a cooperative act that benefits species B, and the return benefit from B does not come directly to the initially cooperative A individual but instead to other A individuals, can that cooperative behavior increase?

This setup is equivalent to a public goods problem. The initial species A actor effectively releases a factor that is beneficial to all of its species A neighbors that can receive the return benefit from B.

If the A recipients are related to the initial A actor, then the cooperative behavior can increase. In this case, species B is a passive reflector of the behavior, and the neighborhood comprising z is the group of A individuals that may act cooperatively and receive return benefits. We could refine that a bit. But typically what matters is that the A actor and the A recipients have sufficiently positive relatedness, r.

What processes create trait correlations between species? If species B varies phenotypically in the return benefit provided to species A, then trait correlations between species can matter.[113,116,119] A correlation may, for example, be between the greater than average tendency of local A individuals to provide a benefit to nearby B individuals and the greater than average tendency of those nearby B recipients to provide a return benefit to the original A cooperators.

What causes such correlations between species? Familial identity by descent is out because kinship does not occur between species. Instead, spatial associations likely arise by physical linkage or codispersal between individuals of different species.

Spatial linkage between species creates genetic associations in the same way that, in classical Mendelian genetics, physical linkage of genes on chromosomes creates linkage disequilibria. Mixing the paired members of different species by uncorrelated dispersal is similar to chromosomal recombination.

Selection of favored combinations between species also creates associations.[113,116,119] When genetic variants in each of two species work well together and are spatially near each other, the paired individuals reproduce more successfully. Bad gene combinations between species reproduce less.

The expansion of good pairs and loss of bad pairs creates genetic associations between species in the same way that positive epistasis between genetic loci creates linkage disequilibria in Mendelian genetics. Overall, the associations depend on the balance between the enhancement by physical linkage or positively selected pairings and the degradation by uncorrelated dispersal.

How do novel cooperative codependencies between species arise? It could happen in a stepwise process.[171] First, species B acts as a passive reflector of A's behavior. Some genetic variants of A act relatively more cooperatively toward B. The enhanced growth of B returns benefits to individuals of A who are genetically correlated with the initial cooperators.

After this first step, in which the cooperative behavior of A toward B rises to a high level, the reciprocal cooperative behavior of B toward A evolves by the same process. Variants of B enhance A's growth. Those A individuals with enhanced growth return additional benefits to B individuals who are genetically correlated with the cooperative behavioral tendency toward A.

As the mutually beneficial traits become common in each species, the pairs may evolve to depend on each other. Such dependency arises because each species becomes part of the environment of the other species.

Instead of sequential steps, the two processes may overlap because both A and B vary genetically in the tendency to produce traits that enhance the growth of the partner species. In this case, there will be a transient period during which particular pairings between species work well together because the pairs carry mutually enhancing genetic variants.[113]

That positive synergism will create genetic associations between species in cooperative behaviors. Such synergism may allow mutual cooperation and codependency to arise in cases for which such mutualism would not evolve without the extra impetus provided by the genetic correlations between species.

Once strong synergism evolves between species, the genetic variation in cooperative tendency on each side may decline. The genetic correlations act as a transient impetus to push the species over the required threshold for the evolution of mutually beneficial traits and possible codependency. With strong codependency, the mutualism becomes irreversible.[116]

5.3   Tradeoffs and Marginal Values

In the tragedy model, increased competitive success against neighbors reduces the efficiency of resource use. Such tradeoffs between fitness components often occur.

Typically, the maximum success arises when the marginal gains between the alternative fitness components become equal. Suppose, for example, that two fitness components trade off against each other. If investing a little more in one component provides a gain that is greater than the loss for investing a little less in the other component, then it pays to shift investment toward the first component.

A maximum occurs only when marginally shifting investment between the two components does not alter overall success. In other words, the marginal changes for each component must be the same.

For example, consider a tragedy model in which individual competitiveness against neighbors depends on the resource uptake rate, y. The average resource uptake rate in the local group is z. Let the focal individual's share of local group success be $I(y, z)$, and the group efficiency in using resources be $G(z)$. Individual fitness is[122]

$$w = I(y, z)G(z),$$

the product of the individual's competitive share of group success, I, and the overall group's success, G.

In the prior section, $I = y/z$ and $G = 1 - z$. Here, we consider the more general functional forms, which may include nonlinear relations.

Normalizing fitness to be one at the evolutionarily favored trait value often helps to obtain a consistent interpretation of forces. Writing

$$w = \frac{I(y, z)}{I(z^*, z^*)} \frac{G(z)}{G(z^*)}$$

yields a normalized fitness of one when evaluated at the fixed point, $y = z = z^*$.

We obtain the trait value that maximizes fitness by following the steps in eqn 5.7, yielding

$$\frac{dw}{dg} = \frac{I_y}{I} + r\left(\frac{I_z}{I} + \frac{G_z}{G}\right) = 0,$$

in which a subscript means a partial derivative with respect to that variable. All functions are evaluated at the fixed point. Matching marginal costs and benefits to eqn 5.3 yields

$$-C_m = \frac{I_y}{I} \qquad B_m = \frac{I_z}{I} + \frac{G_z}{G}.$$

We weight the marginal benefits by r to measure the marginal valuation with respect to the focal individual's fitness. The marginal costs and benefits equalize at the maximum, yielding

$$C_m = r B_m.$$

A simple example of nonlinearity arises when $I = y/z$ and $G = 1 - z^s$ for $0 < s \le 1$. This example describes the earlier tragedy model but with more rapid degradation of group success, G, as average group competitiveness, z, rises from zero. With these assumptions

$$z^* = \left[\frac{1 - r}{1 - r(1 - s)} \right]^{1/s}.$$

When $s = 1$, we recover $z^* = 1 - r$ in eqn 5.4. As s declines, the degradation in group success rises at an increasing rate as group competitiveness, z, increases from zero. That greater loss in group efficiency for small increases in competitiveness reduces the favored level of competitiveness and, equivalently, increases the favored level of cooperation.

5.4   Repression of Competition

Competition degrades efficiency. In a competitive group, everyone's success may decline. All would do better if some mechanism repressed competition.[7,123] Repression of competition is sometimes referred to as *policing*[115,336] or *cheater control.*[411]

Consider a simple extension of the tragedy model in eqn 5.1,

$$w = (1 - c\alpha)\left(a + (1 - a)\frac{y}{z} \right)(1 - (1 - a)z).$$

Individual and average group competitiveness, y and z, remain the same. The trait α is an individual's investment in repressing competition between group members. The group average for repression of competition is a. The level of repression in the group varies between $a = 0$ for free competition and $a = 1$ for complete repression of competition.

In the first term, c is the cost to an individual for investment in policing competition.

In the second term, an individual's competitive success in the group depends on the fraction of resources divided fairly in the absence of competition a, plus the fraction of resources divided under competition, $1 - a$, multiplied by the relative success of the focal individual in competition, y/z.

In the third term, group success degrades in proportion to the fraction of resources allocated by open competition, $1 - a$, multiplied by the average competitive level of group members, z.

Following the prior section, we assume that all variation vanishes at the maximum of fitness, so that $y = z = z^*$ and $\alpha = a = a^*$. We find those maximum values by evaluating how fitness changes with individual competitiveness, y, and investment in policing to repress competition within the group, α. The derivatives $dw/dg_y = 0$ and $dw/dg_\alpha = 0$ express the changes in fitness with the genetic values for the traits.[115,123] Let the similarity coefficients be the same for the two traits, $r = dz/dg_y = da/dg_\alpha$.

When $r > 1 - c$, investing in policing to repress competition does not provide sufficient benefit to individuals, and $a^* = 0$. With no policing, competitiveness rises to the tragedy of the commons value $z^* = 1 - r$, as in eqn 5.4.

We can write the condition equivalently as $c > 1 - r$. We then see that when similarity, r, is sufficiently high, the amount of competition, $z^* = 1 - r$, that could be repressed and provide gains for policing falls below the cost of policing, c. Thus, strong similarity and an intrinsic tendency to cooperate disfavor repressing competition because there is relatively little intrinsic competitive tendency to repress.

As similarity declines, the tendency to compete rises. With more competition, the potential gains for repressing competition increase. Figure 2 of Frank[123] shows the quantitative analysis of this model. The joint evolution of policing mechanisms that repress competition, a^*, and competitiveness, z^*, respond in interesting ways to changes in the costs of policing and competitiveness.

Comparatively, mechanisms that repress competition tend to be more strongly favored as the similarity between neighbors declines.[115] Similarity by itself favors self-restraint and reduced competitiveness. Thus,

$$r \rightarrow \text{self-restraint} \dashv \text{repression of competition.}$$

Many articles discuss repression of competition in microbes.[397,411,439] Despite the potentially powerful force favoring repression of competition, it remains unclear how often such mechanisms occur in nature.

5.5   Heterogeneity in Vigor and Public Goods

Public goods arise when an individual bears the cost for a trait and all group members share equally in the gains.

Repression of competition provides a public good. Individuals pay the cost to repress competition. Group members share the gains for reduced competitiveness and increased efficiency. Similarly, secreted molecules also provide public goods. Secreting individuals bear the cost of production. All neighbors share the benefits.

The costs for producing a public good may vary between individuals. Some individuals may be more vigorous or have access to greater resources. The relative cost to an individual for expressing a public good declines as vigor increases. More vigorous individuals may be more likely to express public goods because of their lower relative costs.[118] This section summarizes the models in Frank.[127]

No Heterogeneity

We first establish the basic setup without heterogeneity between individuals. Let fitness be

$$w = \left[\frac{1 - c(y)}{1 - c(z^*)} \right] \frac{b(z)}{b(z^*)}. \tag{5.8}$$

Individual production of the public good, y, reduces the direct individual component of fitness by the cost, $c(y)$. We normalize the individual fitness component by $1 - c(z^*)$ to get a meaningful scale for costs, where z^* is the average of y across all groups in the population.

The average of individual contributions to public goods within the focal group is z. The group's public goods provide a benefit to individual fitness by the group efficiency term, $b(z)$. We normalize the benefit by the population average value, $b(z^*)$.

We evaluate $dw/dg = 0$ at $y = z = z^*$ to find the trait favored by selection, as in prior sections. The marginal costs and benefits equilibrate at $C_m = r B_m$, yielding

$$\frac{c'}{1 - c} = r \frac{b'}{b},$$

in which primes denote the slopes of each function obtained by differentiation. For linear costs and benefits, $c(y) = y$ and $b(z) = z$. When evaluated at $y = z = z^*$, we obtain

$$z^* = \frac{r}{1+r}.$$

This result differs from eqn 5.5 because I switched from considering z as competitiveness in the prior model to considering z as cooperative public goods production in this model. I switched notation here to match the analysis in Frank,[127] which the following subsections summarize.

BASELINE SUCCESS AND STARTUP COSTS

There may be some productivity in the absence of the public good. For the linear case, we may write benefits as $b(z) = s + z$, so that there is a fixed productivity of s in the absence of the public good.

Producing a public good may require turning on a complex pathway. Making a low level of a public good may be significantly costly because of the startup costs of production. Increasing production from low levels may not add much additional expense. For the linear case, assume that $c(y) = k + y$ for $y > 0$ and $c(0) = 0$, in which k is the startup cost for producing the public good.

Using these benefit and cost assumptions in eqn 5.8, we obtain

$$z^* = \frac{r(1-k) - s}{1+r}. \tag{5.9}$$

Higher baseline success, s, and startup costs, k, reduce production of public goods.

INDIVIDUAL HETEROGENEITY

Suppose individuals divide into classes, j, with resource or vigor level, $1 + \delta_j$, such that δ_j describes the class deviation in vigor from the central value of one. Then we can write individual fitnesses as

$$w_j = \left[\frac{1 + \delta_j - c(y_j)}{1 + \delta_j - c(z_j^*)} \right] \frac{b(z)}{b(z^*)},$$

in which y_j is the contribution to public goods for a focal individual in class j, and z_j^* is the optimal value for class j individuals at equilibrium.

The values of z and z^* are the group average and population average values of the trait.

If we assume linear costs and benefits with baseline success, s, and startup costs, k, as in the prior subsection, then following our usual methods and the details in Frank,[127] we obtain

$$z_j^* = 1 + \delta_j - k - \frac{s + z^*}{r},$$

in which z^* is given by eqn 5.9. If the parameters satisfy $z_j^* \geq 0$ for all of the classes, j, and we assume a symmetric distribution centered at zero for deviations in resources or vigor, δ_j, then

$$z_j^* - z^* = \delta_j.$$

Class j individuals deviate in their public goods expression from the central value of z^* by δ_j, their deviation in vigor from the average.

Comparatively, we obtain the two predictions given in Section 4.4 for heterogeneity. First,

$$\text{vigor} \dashv \text{marginal costs} \dashv \text{secretion.}$$

Increasing vigor reduces the marginal costs of production, which favors greater production of public goods. Marginal costs decline with vigor because, as δ_j rises, a small change in costs, c, has proportionately less effect on baseline individual fitness, $1 + \delta_j - c$. Second,

$$\text{het vigor} \to \text{het costs} \to \text{het secretion.}$$

Increasing heterogeneity (het) in vigor increases heterogeneity in the marginal costs of production, which favors greater heterogeneity in the production of public goods.

5.6   Demography and Reproductive Value

A trait often influences different components of fitness. For example, faster growth of a pathogen within a host increases the pathogen's number of progeny and the dispersal to other hosts. Faster pathogen growth may also decrease the host's lifespan, reducing the survival of the pathogen.

To study microbial traits that trade off dispersal versus survival, we must consider the relative valuation of those two distinct fitness components. Life history theory analyzes the reproductive values of different fitness components.[61,403]

Similarity selection often affects the various reproductive value components of fitness in different ways. Thus, we need to combine the life history analysis of reproductive value with the analysis of similarity.[122,405]

This section briefly illustrates the main concepts. Chapter 8 of Frank[122] provides details, extending Taylor & Frank's[405] original analysis.

PRINCIPLES

Different fitness components associate with different classes of reproduction. For example, we may label dispersers as class 1 individuals and nondispersers as class 2 individuals. The transmission of trait values to the future flows separately through the two classes.

We wish to study the total fitness effect caused by a change in trait value. To obtain the total effect, we analyze the consequences for each class and then combine the results into an overall effect.

The fitness consequence for each class depends on that class's contribution to the future population. The contribution has three aspects.

First, the number of individuals in class j influences the contribution of that class. We write u_j for the frequency of class j.

Second, when class j individuals contribute to class i, the value of that contribution must be weighted by the reproductive value of class i, written as v_i.

For example, if class i represents dispersing individuals, then we must weight the contribution to class i by the expected relative contribution of a disperser to the future population.

Third, the relative contribution of class j to class i is w_{ij}. For example, we may be interested in $w_{ij}(y, z)$, expressing the effect of an individual's trait, y, and the group average trait, z, on the contribution from j to i.

The overall fitness valuation for the contribution of class j to class i is $v_i w_{ij} u_j$. Summing all transitions yields

$$W = \sum_{ij} v_i w_{ij} u_j = \mathbf{v} \mathbf{A} \mathbf{u}. \tag{5.10}$$

Here, $\mathbf{v}$ is the row vector of reproductive values per individual for each class, $\mathbf{u}$ is the column vector of class frequencies, and $\mathbf{A}$ is the matrix of w_{ij} fitness values.

We can study the direction of change in traits and find trait values that maximize fitness by analyzing dW/dg, as in earlier sections. The extended method here accounts for the different numbers of individuals in various classes and the different reproductive valuations for various components of fitness.

DISPERSAL VERSUS SURVIVAL

Consider a tradeoff between the production of dispersing progeny and the future survival in the current habitat. This brief summary follows the model in section 8.3 of Frank.[122]

This example has two classes. Dispersers that successfully colonize a new patch form class 1. The new colonizers and their nondispersing descendants form class 2. Let the fitness components be

$$\mathbf{A} = \begin{bmatrix} 0 & \beta(y)/D \\ 1-t & 1-\delta-z \end{bmatrix}. \tag{5.11}$$

Entries in row i and column j denote w_{ij}, the contribution of class j individuals to class i. Thus, w_{11} is zero because newly arrived colonizers of class 1 do not make dispersers but instead survive locally at rate $w_{21} = 1-t$ to form the surviving lineage of colonizers as class 2.

The component $w_{12} = \beta(y)/D$ describes the contribution of the local lineage to dispersers that successfully colonize a new patch. The local lineage's investment in making dispersers is y, and $\beta(y)$ is the functional relation between dispersal investment and dispersal success. Dispersal success is normalized by the density-dependent factor, D, in which greater density-dependent limitation reduces dispersal success.

The component $w_{22} = 1-\delta-z$ describes the survival of the colonizing lineage within its patch. The intrinsic loss rate is δ, which combines destruction of the patch, loss of the colonizers from a continuing resource patch, or death of a host when the colonizers are parasites.

The intrinsic loss rate is increased by z, which is the patch average of the trait value y that determines the number of successful dispersers. As successful dispersal rises, the local survival rate decreases.

When evaluating total fitness, W, from eqn 5.10, we need the individual reproductive values, $\mathbf{v}$, for the classes when evaluated at demographic equilibrium, $y = z = z^*$, derived in Frank[122] as

$$\mathbf{v} \propto \begin{bmatrix} 1-t & \lambda \end{bmatrix},$$

in which "∝" means *proportional to.* The reproductive value of new colonizers is discounted by $1 - t$, the probability of surviving the initial delay after colonization and before producing dispersers. The reproductive value of residents is augmented by λ, the population growth rate, because residents have average reproductive success λ during the period when new colonizers do not reproduce. The value of λ is the dominant eigenvalue of the fitness matrix $\mathbf{A}$ evaluated at $y = z = z^*$.

The class frequencies at demographic equilibrium are proportional to

$$\mathbf{u} \propto \begin{bmatrix} \beta(z^*)/D \\ \lambda \end{bmatrix}.$$

To obtain the trait values that maximize the total fitness in eqn 5.10, we evaluate $dW/dg = 0$ at $y = z = z^*$, which includes

$$\frac{d\mathbf{A}}{dg} = \begin{bmatrix} 0 & \beta'(z^*)/D \\ 0 & -r \end{bmatrix}$$

and the vectors $\mathbf{v}$ and $\mathbf{u}$ at demographic equilibrium, leading to a solution that must satisfy $v_1 \beta'(z^*)/D = v_2 r$, which yields

$$\beta'(z^*) = \frac{r\lambda D}{1 - t}.$$

If we assume that dispersal success is $\beta(z) = z^s$, with $s < 1$, then dispersal success rises at a diminishing rate with investment in dispersal, yielding the solution

$$z^* = \left[\frac{s(1-t)}{r\lambda D} \right]^{1/1-s}. \tag{5.12}$$

The various terms interact to determine the favored dispersal rate, z^*. However, we can get a sense of partial causation by considering how z^* changes in response to partial changes in the terms. In particular, a rise in t lowers the initial survival of colonizers within a patch, decreasing investment in dispersal. Similarly, a rise in λ raises the growth of patch residents, lowering the relative value of colonizers and also decreasing investment in dispersal.

A decrease in density-dependent limitation, D, increases the opportunity for dispersers to find new patches, raising dispersal. Smaller values of s cause more rapid saturation of dispersal success, lowering dispersal investment.

This model also expresses the tragedy of the commons. Reduced similarity, r, favors more dispersal, which decreases local survival and the long-term quality of the local patch. In other words, dispersal is a competitive trait that degrades the local commons by more rapidly extracting local resources to develop dispersal-enhancing traits.

These conclusions provide a rough qualitative sense of how various forces shape dispersal. In each case, I emphasized how a change in some factor leads to a partial pathway of causation favoring either an increase or a decrease in dispersal.

ALTERNATIVE MECHANISTIC EFFECTS ON DISPERSAL AND SURVIVAL

In the fitness matrix of eqn 5.11, individual trait value, y, influences dispersal, and the group average trait, z, influences local survival. These assumptions express a tragedy type model, in which individuals compete for resources to increase the dispersal of their progeny, and competitiveness degrades the local commons.

Alternatively, successful dispersal may require joint action by neighbors, so that dispersal depends on the group average trait, z. Each individual's cooperative contribution to joint action, y, reduces its own survival but does not affect the survival of neighbors. These assumptions create a public goods problem. Individual traits contribute to dispersal, which arises from shared public goods.

In the prior model, we change from the original tragedy assumptions to the public goods problem by switching y and z in the fitness matrix of eqn 5.11. We then obtain the same form for the favored trait value as in eqn 5.12, but with the similarity coefficient, r, now in the numerator rather than the denominator

$$z^* = \left[\frac{rs(1-t)}{\lambda D} \right]^{1/1-s}.$$

In this case, increasing similarity favors greater cooperative contribution to dispersal as a public good, with a greater individual cost through lower survival.

These examples show how alternative mechanistic aspects of traits can reverse the direction of trait evolution favored by a particular force.

THE CENTRAL ROLE OF DEMOGRAPHY

Under different assumptions, a rise in population growth rate may provide better opportunities for dispersal rather than better success for residents. In that case, increasing population growth rate, λ, would associate with greater investment in dispersal.

The point is that demographic processes can strongly influence the direction of trait evolution. In studies of microbes, past work has emphasized similarity and kin selection but has paid relatively little attention to demographic aspects of populations.

5.7 Stage-Dependent Traits in Life Cycle

Suppose some microbes colonize a resource patch. They grow for many generations. They also send dispersers to colonize other patches. Those dispersers can be thought of as the reproduction or fecundity of the group. Total reproduction over the colony life cycle depends on how long the colony survives.

Over the colony life cycle, how do the fundamental forces shape competitive and cooperative traits? We must consider, at each stage in the life cycle, how traits influence an individual's relative share of the group's current and future genetic transmission. We must multiply that reproductive share by the total productivity of the group.

We could use the demographic methods of the previous section to analyze the various components of fitness. However, it is easier in this case to write a single expression that combines the fecundity and survival components of fitness over the full life cycle. This section briefly summarizes Frank's[128] analysis.

CYCLE FITNESS

A colony grows through $j = 0, 1, \ldots$ temporal stages. The fitness of a focal individual in the jth stage is

$$w_j = I(y_j, z_j) \sum_{k=j}^{\infty} \lambda^{-k} G(z_k). \tag{5.13}$$

The first term, I, describes an individual's share of the colony's long-term success. In a tragedy model, I increases with an individual's competitive trait expression, y_j. For example, $I = y_j/z_j$ expresses the relative competitive success of an individual with trait y_j when competing in a group with average competitive trait value, z_j. In a public goods model, I decreases with greater individual expression of the public good, y_j.

The second term describes the reproductive value for the colony in the jth stage. That value is the sum of the colony success, G, in the current stage, j, and in all future stages. The colony success for each stage is multiplied by the discount for the amount the population size has grown, λ^{-k}, since colony inception at stage $j = 0$. We discount future reproduction by the expansion of the population size because a single progeny represents a declining share in an expanding population.

The group success in stage k can be divided into survival and fecundity components of reproductive value,

$$G(\mathbf{z}_k) = S(\mathbf{z}_k)F(\mathbf{z}_k).$$

The survival to stage k is $S(\mathbf{z}_k)$, which depends on the group average trait expression in each stage up to and including the current stage, $\mathbf{z}_k = z_0, z_1, \ldots, z_k$. Similarly, the fecundity $F(\mathbf{z}_k)$ also depends on the current and prior trait expression.

We find the trait vector, $\mathbf{z}^*$, that maximizes fitness by simultaneously evaluating $dw_j/dg_j = 0$ for all j when evaluated at $\mathbf{y} = \mathbf{z} = \mathbf{z}^*$.

TRAGEDY AND PUBLIC GOODS MODELS

Suppose the colony grows without producing dispersers from generations $k = 0, 1, \ldots, g - 1$. Then surviving colonies remain at constant size and produce migrants in proportion to their fecundity in each of the following generations.

With those assumptions, the components of individual success, group survival, and group fecundity are, respectively

$$I(y_j, z_j) = \frac{y_j}{z_j}$$

$$S(\mathbf{z}_k) = S(\mathbf{z}_k^*)\left[\frac{1 - z_j}{1 - z_j^*}\right]^{\theta(g-1-j)}$$

$$F(\mathbf{z}_k) = F(\mathbf{z}_k^*)\left[\frac{1 - z_j}{1 - z_j^*}\right].$$

Individual success follows the standard tragedy model. An individual's share of group success in the jth generation is the ratio of its competitive trait, y_j, relative to the group average, z_j.

Survival to generation k depends on the survival in each of the preceding generations. Thus, any cooperative enhancement of survival in a particular generation carries a benefit forward to all future generations. In this model, deviations in group trait values only influence survival during the juvenile generations, $j < g - 1$.

In each juvenile generation, j, the survival consequence of a deviation in group trait value, z_j, is $[(1 - z_j)/(1 - z_j^*)]^\theta$. That value multiplies for each of the $g - 1 - j$ juvenile generations over which it acts. Any consequence to total survival over the juvenile period also affects cumulative survival to future reproductive generations. The value of $S(\mathbf{z}_k^*)$ is the baseline survival rate to generation k in a group without deviant trait values.

The fecundity consequence for a deviation in group trait value is $(1 - z_j)/(1 - z_j^*)$. The value of $F(\mathbf{z}_k^*)$ is the baseline fecundity in generation k in a group without deviant trait values.

This model assumes the typical tragedy of the commons form, in which increasing z corresponds to greater competitiveness and degradation of group success. We can also interpret this model as a public goods problem, in which $1 - y$ is an individual's public goods production and $1 - z$ is the group's average production. Then decreasing z corresponds to greater cooperative contribution to public goods and an increase in group success.

With these alternative model interpretations, we can think of $1 - z_j^* : z_j^*$ as the ratio of the cooperative to competitive tendency in trait values.

Solving $dw_j/dg_j = 0$ for all j when evaluated at $\mathbf{y} = \mathbf{z} = \mathbf{z}^*$ yields z_j^*, the favored trait value in each generation j. When expressed as the cooperative to competitive tendency, $1 - z_j^* : z_j^*$, we obtain

$$r(1 + y_j) : 1 - r,$$

with the enhanced demographic component for the cooperative tendency caused by the trait's contribution to colony survival as

$$y_j = \begin{cases} \theta(g - j - 1) & j < g - 1 \\ 0 & j \geq g - 1. \end{cases}$$

This model illustrates the increased selective force on cooperative and competitive traits during the early stages of colony growth, when j is small. More detailed mechanistic assumptions for trait action would lead to more specific predictions for particular traits. For example, a secreted public good that decays more slowly than the generation time would be strongly favored early in the colony growth cycle but would be less advantageous later in the colony life cycle.

The declining value of new secretions arises in two ways. First, public goods may already be present in the environment because of the slow decay from secretions in prior generations. Second, cooperative traits later in the colony life cycle typically have lower reproductive value.

5.8   The Three Measures of Value

This section briefly summarizes the primary measures of value. Three exchange rates transform the various effects of traits into the common currency of contribution to the future population.[122]

First, interacting individuals may have similar trait values or share similar genes. The coefficient, r, relates the similarity of individuals to the consequences for reproductive success and heritable transmission to the future, the primary currency.

Second, marginal values compare a trait's effects on different components of fitness. The favored trait value, when altered by a small amount, typically causes equal marginal gains and losses between its associated fitness components. If a changed trait caused a larger marginal gain in one component than the offsetting loss in another component, then the trait value would tend to change until it settled near the favored balance of marginal gains and losses.

Third, reproductive value compares a trait's influence on different pathways of heritable transmission to the future population. For example, a trait may influence survival, fecundity, and dispersal. A gain in one component may be offset by a loss in another component. To compare the gains and losses, each component must be expressed in terms of its ultimate contribution to the future population, the component's reproductive value.

We may also assign reproductive values to different life stages or to different kinds of habitat. In each case, the classification can be used to analyze the class's relative contribution to the future population, which is its reproductive value. In tradeoffs, the fundamental forces typically favor traits that positively influence classes with relatively high reproductive value.

5.9   Scaling of Time and Space

Forces acting over short timescales may oppose forces acting over longer time periods. Consider a fast-growing mutant. The mutant outcompetes its neighbors, rising in frequency immediately, over a short timescale.

Faster growth may associate with poor conversion efficiency of food into reproduction. Poor yield typically acts over a longer timescale as resources slowly become depleted (p. 138).

Inefficient resource use may, for example, lower a group's long-term production of dispersers to colonize new habitats. Group against group competition happens more slowly than the direct competition between individuals within groups.[446]

The design of traits depends on the balance between within-group forces acting over short timescales and between-group forces acting over long timescales.[133,168,269,433,447]

The spatial scale of competition influences the relative weighting of different timescales. When the spatial scale of competition is large, and individuals compete globally with each other across all spatial locations, then relative success depends only on the direct and immediate competition between individuals. The short timescale dominates.

When the spatial scale of competition is small, and individuals compete locally, then total success depends on the balance of the two forces. Competition within groups favors fast growth, acting over short timescales. Competition between groups favors high yield, acting over long timescales.

Short timescales act rapidly and ubiquitously. Long timescales act slowly and sporadically. All else equal, the short timescales dominate.[446]

But all else may not be equal.[168] If there is relatively little variation between individuals within groups, then within-group competition has relatively little consequence. The long timescale of between-group competition dominates.

By contrast, if most of the variation occurs between individuals within groups, then not much difference occurs between groups. Limited between-group variation means that there is only a small force of competition at that longer scale. The short timescale of within-group competition dominates.

The next subsection sketches the basic theory for relative variation and timescale. The following subsection considers situations in which competitive and cooperative interactions happen at different spatial scales.

TIMESCALE AND THE PROCESSES THAT GENERATE VARIATION

The force acting at each scale depends on the intensity of selection.[133] Within groups, we may write the opportunity for outcompeting neighbors as the intensity of selection, s_w.

Selection has consequences only when competition occurs between differing individuals. If individuals carry the same genes, then with regard to evolutionary change, it does not matter which one wins in competition. We express the differences between individuals within groups as V_w, the within-group variance.

The evolutionary force within groups scales as $s_w V_w$, the product of the potential for differences in success multiplied by the variance. Between groups, we write $s_b V_b$, the potential for differential success between groups multiplied by the variance between groups.

When the two scales oppose each other, then traits evolve toward a balance between the opposing forces,

$$s_b V_b = -s_w V_w. \tag{5.14}$$

Equality requires changing the sign on one side of the equation because opposing forces have opposite signs.

The variance between groups, V_b, versus the variance within groups, V_w, determines the relative weighting of selection at the global versus local spatial scales.[168,434] We can relate those variances to similarity and kin selection by expressing the values as descriptions of relative similarity. To begin, we write the total variance as

$$V_t = V_w + V_b.$$

We then define the relative similarity of individuals within groups as the fraction of the total variance that is between groups,

$$r = \frac{V_b}{V_t}.$$

The more of the total variance that occurs between groups, the lower the fraction of the total variance that occurs within groups. Less variability within groups is the same as more similarity within groups. Here, r

is the correlation coefficient between individuals within groups, which expresses the coefficient of similarity within groups.

Substituting those identities for similarity and variance into eqn 5.14, the balance of opposing forces occurs when

$$s_b r = -s_w (1 - r). \tag{5.15}$$

As similarity within groups, r, increases, the force, $s_b r$, between groups rises and the force, $s_w (1 - r)$, within groups declines. The balance shifts toward selection between groups. Thus, as r rises, selection increasingly favors traits that enhance competition between groups, often reducing competition or raising cooperation within groups.[112]

The basic tragedy of the commons model follows directly from the balance of forces. Group success in the basic tragedy model of eqn 5.1 is $1 - z$, and thus $s_b = -1$, the slope of group success with respect to the average trait value in groups, z. The selective intensity within groups is $s_w = (1 - z)/z$, which is the partial change in individual fitness, w, with respect to the change in individual character value, y, holding constant group phenotype, z. Substituting these values for selective intensity into eqn 5.15 and evaluating at the fixed point z^* yields the basic result for the tragedy, $z^* = 1 - r$, given in eqn 5.4.

The distinction in this section arises from a focus on the relative timescales for the different forces and a clearer spatial separation of processes within and between groups. With those explicit considerations of dynamics, we get a better sense of the forces that shape traits.

For example, the previous analyses of similarity selection took r as a given value. But what, in fact, determines the value of r?

If the generation of variation happens slowly, on a long timescale compared with selection, then the distribution of variation within and between groups arises by the way in which individuals assort spatially. Common ancestry is often the most powerful cause of spatial assortment and similarity within groups. In that case, similarity and the associated value of r arise by kinship, leading to the natural interpretation of similarity selection as kin selection.

In many multicellular organisms, new variation arises slowly. Most aspects of similarity depend on kinship. Other factors may sort similar individuals into groups, but kinship typically dominates.

Microbes differ. Short generation times and large population sizes mean that mutation and selection within groups can create new variation relatively rapidly compared to spatial sorting by common descent. Thus, the processes that generate the distribution of variation may happen on the same timescale as selection. Kinship and common descent do not necessarily dominate the spatial patterns of similarity and variance.

The generation of new variation within groups degrades the local similarity and decreases r over time. The decay of within-group similarity shifts the balance of forces toward within-group competition, reducing the potential for within-group cooperation.

The dynamic changes of force that increase within-group competition may lead to microbial cancers, in which highly competitive variants arise and overgrow their neighbors, degrading the long-term success of groups (Section 4.3).

In microbes, different species often strongly interact over short spatial and temporal scales. Similarity selection occurs between species when processes other than kinship cause similarity in trait values.[113,116,119]

SPATIAL SCALE OF COMPETITION VERSUS COOPERATION

Competitive and cooperative interactions may happen over different spatial scales. For example, secreted public goods may act locally, cooperatively benefiting only close neighbors. By contrast, key resources that competitively limit growth may diffuse over relatively longer spatial scales.

Those different spatial scales influence the costs and benefits that shape cooperative traits. Consider the expression of fitness from section 7.1 in Frank,[122]

$$w = \frac{bz - cy}{az(b - c) + (1 - a)\bar{z}(b - c)},$$

in which an individual invests y in cooperative public good secretion, at cost cy to itself. The average level of altruistic public goods expression in the neighborhood is z, with beneficial effect bz on fitness. The focal individual's reproduction is therefore proportional to $bz - cy$, which is the numerator.

The denominator is the intensity of competition for scarce resources. Competition increases as the average reproductive success rises. The overall level of competition combines local and global components of reproductive competition.

The average local reproduction in the neighborhood is the average of $bz - cy$, which is $z(b - c)$ because the local average of y is z. The average in the population is $\bar{z}(b - c)$. The parameter a is the spatial scale of density-dependent competition. An increase in the reproductive success of neighbors by a proportion δ increases local competition by a factor $a\delta$. An increase in the average reproductive success of the population by a proportion y increases global competition by a factor $(1 - a)y$.

Using our standard method in eqn 5.7 to find the trait value favored by natural selection, we obtain the condition for the cooperative trait to increase,[122]

$$\frac{dw}{dg} = r[b - a(b - c)] - c > 0,$$

in which the marginal benefit is $B_m = b - a(b - c)$, and the marginal cost is $C_m = c$.

Comparatively, as a rises and competition for resources becomes increasingly local, the tendency for cooperative trait expression declines. Local competition reduces the benefit of cooperative traits because an increase in neighbors' vigor from enhanced cooperative expression is offset by the increased competition among those more vigorous individuals for the same locally limiting resources. Put another way, limited local resources reduce the potential for enhanced success through cooperative traits.[6,112,333,404,449]

In general, a trait may alter various components of fitness acting at different spatial scales. The changed balance of forces at the various scales modifies the trait's design.

5.10 Variable Environments

Reproduction multiplies. If a population grows by μ in each generation, the total population growth is $\mu \times \mu = \mu^2$ after two rounds of reproduction.

Variation in reproduction lowers fitness. For the same average population growth rate of μ, if the rate goes up by δ in the first generation and down by the same amount in the second generation, then total growth is $(\mu + \delta)(\mu - \delta) = \mu^2 - \delta^2$. Variation in reproduction reduces success.[99,154,176,223,304,331,341,414]

The multiplicative nature of reproduction leads to the geometric mean principle. The next subsection discusses the geometric mean, which shows how variation discounts value.

The current literature emphasizes the geometric mean principle but mostly ignores other aspects of variation that influence value. We obtain a deeper sense of biological design by thinking about what traits do in organismal life history and how different kinds of variation alter value.

After introducing the geometric mean, the following subsections show other ways in which variation influences value.[131,144,153,234,274,306,337]

GEOMETRIC MEAN

The total growth after t generations is the product of the growth, λ_i, in each generation,

$$\Lambda = \prod_{i=1}^{t} \lambda_i.$$

Because growth multiplies, there must be some value, λ, that we can multiply with itself t times to get the same total growth. In symbols, t multiplications of λ is λ^t. Thus, we can write $\lambda^t = \Lambda$ as the total growth and then figure out what sort of average value λ is.

Taking the natural logarithm of both sides yields the same equality, now written as a sum of logarithms on the right-hand side,

$$\log(\lambda^t) = \log\left[\prod_i \lambda_i\right] = \sum_i \log(\lambda_i).$$

Define $m = \log(\lambda)$ and note that $\log(\lambda^t) = t\log(\lambda) = tm$. Then

$$m = \frac{1}{t}\sum_i \log(\lambda_i)$$

is the average of the logarithmic growth rate, the Malthusian parameter. The geometric mean is defined as $\lambda = e^m$. Total growth is t multiplications of the geometric mean growth rate,

$$\lambda^t = e^{mt}.$$

Variation in the individual growth rates per generation reduces the geometric mean and the total growth. As noted above, if we let μ be the arithmetic mean growth rate per generation and suppose, over two generations, that growth fluctuates up and down by δ, then

$$\lambda^2 = (\mu + \delta)(\mu - \delta) = \mu^2 - \delta^2.$$

Increasing fluctuation, δ, always reduces the total growth.

No simple mathematical expression describes exactly how increasing variation causes a greater discount to the total growth. For arithmetic mean and variance in growth rate per generation, μ and σ^2, the geometric mean is approximately

$$\lambda \approx \mu - \sigma^2/2\mu.$$

The smaller the variance relative to the mean, the better the approximation will be. Because the arithmetic mean growth rate is often near one in evolutionary models, $\mu \approx 1$, the geometric mean approximation is often written as

$$\lambda \approx \mu - \sigma^2/2.$$

In summary, the geometric mean measures the long-term growth rate. Increased arithmetic mean growth per generation, μ, may lead to a lower geometric mean fitness value if the enhanced growth also causes a sufficiently large increase in the variance, σ^2. In general, variation in performance discounts long-term value. Natural selection often favors traits that reduce variation in performance.

Absolute versus Relative Success

The geometric mean has often been claimed as a one-step principle for calculating the valuation discount caused by trait variability. However, other aspects also influence the relation between variability and value.[131,144]

Consider the distinction between absolute and relative success. The geometric mean calculates total growth, a measure of absolute success. Relative success is what matters in biology. If one genotype increases tenfold, that increase describes significant success. However, if a competitor increases 100–fold, then the original type has greatly declined in frequency.

The traits that dominate the observable patterns of nature associate with greater relative success. To describe how relative success affects the relation between variability and the discount in value, I summarize Frank & Slatkin's[144] extension of Gillespie's[153] analysis.[131,306]

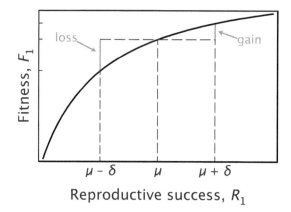

Figure 5.1 Increasing variation in reproductive success reduces fitness, from eqn 5.16. The fitness loss from negative fluctuations outweighs the fitness gain from positive fluctuations. Thus, equally frequent negative and positive fluctuations return a net loss. Redrawn from Frank & Slatkin.[144]

Consider alternative genotypes that encode different trait values. Let q_1 and q_2 be the frequencies of the alternative genotypes. After one round of reproduction, the updated frequency for the first type is

$$q_1' = q_1(R_1/\bar{R}) = q_1 F_1,$$

in which R_1 is the reproductive success or absolute fitness of the first type, $\bar{R} = q_1 R_1 + q_2 R_2$ is the average reproductive success of the two types, and F_1 is the relative fitness of the first type. This equation emphasizes that relative fitness is what controls frequency change and the evolution of traits.

Writing out the definition of relative fitness explicitly in terms of frequency and absolute reproductive success yields

$$F_1 = R_1/\bar{R} = \frac{R_1}{q_1 R_1 + q_2 R_2}. \tag{5.16}$$

A gain in absolute success causes a smaller ultimate benefit than the loss imposed by an equal and opposite decline in absolute success (Fig. 5.1). Put another way, diminishing return causes variability in absolute success to impose a discount on relative success.

The curvature between absolute and relative success depends on frequency (Fig. 5.2). A rare type has a nearly linear relation between

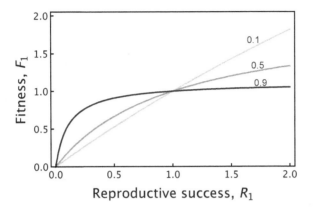

Figure 5.2 The curvature of relative fitness versus reproductive success depends on frequency. The numbers above each curve show q_1. Rising frequency increases the curvature between absolute and relative success. Greater curvature leads to a bigger fitness value discount. The curve for $q_1 = 0.1$ shows that there is little curvature when a type is rare, providing an advantage for rare types. Based on eqn 5.16, with $R_2 = 1$. Redrawn from Frank & Slatkin.[144]

reproduction and relative fitness. A common type has a strongly diminishing relation between reproduction and relative fitness.

More strongly diminishing returns cause variability in reproductive success to impose a greater penalty on relative fitness. Thus, common types, with more strongly diminishing returns between absolute and relative success, suffer a greater discount than do rare types. In general, the variability discount to relative fitness is frequency dependent.[144,274]

Competition for resources makes relative success particularly important. Over time, one cannot simply multiply the reproductive successes of each type independently and then compare the long-term geometric means. Instead, each bout of density-dependent competition causes interactions between alternative types.

The fitness measure of relative success in eqn 5.16 accounts for density-dependent interactions. But that equation does not tell us the temporal and spatial scales over which density-dependent competition acts. Different scalings of competition alter the relation between trait variability and relative fitness value.

Competitive scale varies widely among microbes. How does a change in competitive scale alter the relation between trait variability and fitness? The theory for that question has not been developed in a general way.

EXPECTED CHANGE IN FREQUENCY

The previous subsection analyzed the frequencies of two competing types, q_1 and q_2. The updated frequency of q_1 after a round of competition is $q_1' = q_1 F_1$, in which F_1 is the relative fitness of that type. Thus, $\Delta q_1 = q_1' - q_1 = q_1(F_1 - 1)$. Using the definition of relative fitness in eqn 5.16, we can write the change in frequency for the first type as

$$\Delta q_1 = q_1 q_2 \left(\frac{R_1 - R_2}{\bar{R}} \right).$$

The reproductive successes fluctuate randomly. If the fluctuations in success are small relative to the average success and we normalize the success values to be close to one, then the approximate expected change in frequency is[131,144,153]

$$E(\Delta q_1) \approx q_1 q_2 \{ (\mu_1 - \mu_2) + [\text{cov}(R_2, \bar{R}) - \text{cov}(R_1, \bar{R})] \}, \qquad (5.17)$$

in which μ_1 and μ_2 are the expected reproductive successes for types 1 and 2.

HIERARCHICAL STRUCTURE OF VARIABILITY

Suppose the variance in success for an individual of genotype 1 is σ_1^2. Then the variance in the genotypic success is $\text{var}(R_1) = \rho_1 \sigma_1^2$, in which ρ_1 is the correlation between randomly chosen individuals of that genotype.

When all individuals have the same success, $\rho = 1$, then individual and genotypic variance are the same. As individuals become less correlated, the genotypic variance declines because the variance of an average decreases with the number of uncorrelated samples. We may also write $\text{var}(R_2) = \rho_2 \sigma_2^2$ for type 2, and $\text{cov}(R_1, R_2) = \rho_{12} \sigma_1 \sigma_2$, in which ρ_{12} is the correlation between randomly chosen individuals of types 1 and 2.

If, for simplicity, we assume $\rho_{12} = 0$, then eqn 5.17 becomes[131,144]

$$E(\Delta q_1) \approx q_1 q_2 \{ (\mu_1 - q_1 \rho_1 \sigma_1^2) - (\mu_2 - q_2 \rho_2 \sigma_2^2) \}. \qquad (5.18)$$

On average, type 1 increases in frequency when

$$\mu_1 - q_1 \rho_1 \sigma_1^2 > \mu_2 - q_2 \rho_2 \sigma_2^2. \qquad (5.19)$$

Rare types, with smaller q, gain an advantage. That rare-type advantage occurs because the curvature between reproductive success and fitness increases with frequency (Fig. 5.2), making common types more sensitive to the fitness value discount for variability in reproductive success.

The rare-type advantage tends to push frequencies away from zero, favoring a mixture of types. However, stochastic fluctuations in frequency often cause loss of one of the types, leading to fixation of the other type.

Over time, the frequencies tend to be biased toward the type with the greater geometric mean fitness. That long-term bias can most easily be seen by starting with equal frequencies, $q_1 = q_2 = 1/2$. At that frequency midpoint, type 1 tends to be favored when

$$\mu_1 - \rho_1 \sigma_1^2/2 > \mu_2 - \rho_2 \sigma_2^2/2.$$

This expression compares the geometric means of the two types. A type can potentially lower its overall variance, $\rho_i \sigma_i^2$, and increase its success by reducing the correlation between individuals, ρ_i.

BET-HEDGING

Reducing the correlation between individuals is one type of bet-hedging. For example, if individuals stochastically express alternative traits, then the genotype increases the chance that a subset of individuals match the current state of a varying environment. In general, bet-hedging strategies tend to reduce the overall variance of a genotype's success.[159,400]

SPATIAL SCALE OF COMPETITION

In the previous subsections, competition occurs in one large population. This subsection considers a population distributed over many independent spatial locations. Competition happens within each separate location.

Temporal fluctuations within each location induce frequency dependence, favoring the rare type (eqn 5.19). When there is only a single location, one of the types typically becomes fixed after a period of time because the random fluctuations in frequency are too strong relative to the directional tendency of evolutionary change. Fixation is biased toward the type with the highest geometric mean.[80]

By contrast, in a population distributed over many separate locations, the rare-type advantage typically maintains a mixture of types. The tendency for mixture arises in the following way.[234]

In each time period of local competition, the rare types gain on average in each patch because of their intrinsic frequency-dependent advantage. The population-wide fluctuations in each round of local competition become small because of the averaging effect over the many patches. We can therefore treat eqn 5.18 as an essentially deterministic process. The rare-type frequency dependence now dominates. The equilibrium frequency of types can be obtained from eqn 5.18 by solving $E(\Delta q_1) = 0$, which yields[144]

$$\frac{q_1}{q_2} = \frac{\mu_1 - \mu_2 + \rho_2 \sigma_2^2}{\mu_2 - \mu_1 + \rho_1 \sigma_1^2}.$$

Each ρ is the correlation between individuals of a type measured within each patch. This result shows that geometric mean success does not always provide the correct fitness value.

RELATION BETWEEN TRAITS AND VARIABLE PERFORMANCE

The previous subsections assumed that an individual's variability in reproductive success is a given parameter. This subsection briefly summarizes how an individual's multiple traits combine to determine its overall variability in performance. See the details and examples in Frank.[131]

We begin with a single trait for resource acquisition, in which reproductive success is

$$R = 1 + f(\delta).$$

Random fluctuations in resource acquisition, δ, with mean zero and variance, V_x, affect reproductive success by $f(\delta)$. If fluctuations are relatively small, then the approximate average reproductive success is

$$\mu \approx 1 + f'' V_x / 2,$$

in which f'' is the second derivative of f evaluated at zero.[337] Typically, $f'' < 0$ because the benefits of resource acquisition have diminishing returns. Thus, greater fluctuations, V_x, reduce expected reproductive success. All else equal, resource acquisition strategies with less variability yield higher average reproductive success than those strategies with more variability.

The variance in an individual's reproductive success is approximately

$$\sigma^2 \approx f'^2 V_x,$$

in which f' is the derivative of f evaluated at zero.

To keep the focus on trait variability within individuals, I give only the geometric mean reproductive success for an individual in this subsection. A full analysis of fitness valuation requires the additional aspects discussed in the prior subsections.

An individual's geometric mean reproductive success is approximately

$$G \approx \mu - \sigma^2/2\mu \approx 1 - (f'' - f'^2)V_x/2. \qquad (5.20)$$

Now consider two different traits that provide additive returns. How should an individual divide its investment between those two traits? Assume that reproductive success is

$$R = x[1 + f(\delta)] + y[(1 - y) + g(\epsilon)],$$

in which x and y are the fractions of total resources invested in each trait, y is the small discount in expected return for the second trait, and ϵ is the small random fluctuation associated with the second trait.

Assuming that the fluctuations δ and ϵ are uncorrelated, $V_x = V_y$, and $f \equiv g$, the geometric mean reproductive success for an individual is

$$\mu - \sigma^2/2\mu \approx G + B(x, y),$$

in which G is the geometric mean in eqn 5.20 for allocating all resources to the first trait, $x = 1$, and $B(x, y)$ is the benefit obtained when mixing allocation of resources between the two traits, with $x + y = 1$ and

$$B(x, y) = f'^2[1 - (x^2 + y^2)]V_x/2 - yy.$$

Optimizing B to obtain the best mixture of allocations between the two traits yields

$$x^* = \frac{1}{2}\left(1 + \frac{y}{\sigma^2}\right)$$

$$y^* = \frac{1}{2}\left(1 - \frac{y}{\sigma^2}\right),$$

in which y is the discount in expected return for the second trait, and σ^2 is the variance in individual reproductive success per trait, with $y < \sigma^2$.

It pays to invest some resources in y, the trait with lower expected return. The lower expected return is offset by the benefit from reduced overall variance in performance obtained from averaging the returns over the two uncorrelated traits. This mixed allocation is another type of bet-hedging, the combining of alternative traits to reduce the variation in performance.

In both biology and financial investing, returns tend to multiply over time. Thus, reduced fluctuations enhance the multiplicative (geometric) average return. In financial investing and modern portfolio theory, the geometric mean plays a key role in the allocation of resources among alternative asset classes.[40] In biology, one can think of different traits as different asset classes.

6 Theory: Traits

The prior chapter summarized the forces that change fitness value. Fitness value considers traits abstractly. This chapter reviews how traits develop, what causes traits to vary, and where new traits come from.

The first section contrasts abstract and mechanistic aspects of traits. Abstractly, changed genetic mixing predicts a change in growth rate. That abstract prediction provides broad insight. But it also ignores the mechanistic basis of growth.

Mechanistically, growth depends on the underlying biochemistry and biophysics. The fundamental forces of value shape design through mechanistic change. Mechanistic insight improves the accuracy of comparative predictions and broadens the understanding of design.

The second section discusses the modification of traits. In some cases, small changes to existing traits may be sufficient. Attack less. Disperse more. Heritable quantitative variation often exists, providing the basis to adjust traits.

Big environmental shifts may require large changes in traits. Large variants may not exist. To meet that challenge, the processes that generate variation may evolve. Increased mutation, genomic rearrangement, and genetic mixing generate greater variation. Generative processes modify the evolutionary rate of traits.

The third section considers the origin of traits. How do cells acquire resistance to a novel toxin? How can a cell switch to a novel food source?

In the first step, a novel genotype may arise. However, complex traits often require the simultaneous evolution of several components. A single genetic novelty by itself may be of little value.

Alternatively, the path to a novel trait may begin with a phenotypic variant of a common genotype. The initial phenotypic variant may not produce the favored trait. But it can bring a genotype closer to the favored form.

With a partial solution from an initial phenotypic variant, subsequent genetic variants can more easily transit to a novel character. This phenotypes-first sequence greatly accelerates evolutionary discovery.

Organisms often plastically adjust phenotypes to match the environment. Because plasticity typically covaries several mechanistic components, plasticity may generate variety in the right direction with regard to a novel challenge. Genetic variation and selection can then modulate the initial variety, steadily moving toward the favored form.

6.1 Nature of Traits

More genetic mixing reduces similarity between neighbors, which enhances growth rate to outcompete genetically distinct neighbors. In particular,

$$\text{genetic mixing} \dashv \text{similarity} \dashv \text{growth rate},$$

which expresses an interesting and testable comparative prediction. This prediction considers growth rate abstractly, ignoring the mechanisms that determine the trait.

What determines growth rate? Genes do not encode growth rate. Instead, genes influence the expression of molecules, which alter the uptake of substrates and the sensing of food concentrations. Nucleotide sequences affect the binding kinetics of transcription factors, which trigger switching between metabolic pathways.

Do these mechanisms matter when trying to understand the forces that shape growth rate? At one level, they do not. The comparative prediction for genetic mixing typically holds for different mechanistic assumptions. Not always, but likely often enough that one expects the predicted direction of change in growth to happen more often than not.

At another level, the mechanistic basis of traits provides deep insight into the forces that shape design. Consider two contrasting mechanisms that influence growth rate.

First, a gene duplication may increase the expression of a cell surface transporter that pulls sugar into the cell. Greater uptake rate for sugar may enhance growth rate. Making more transporters requires additional resources, reducing the efficiency yield at which a unit of sugar is transformed into a unit of reproductive biomass. This mechanism creates a tradeoff between growth rate and reproductive yield.

Second, a nucleotide substitution in an enhancer of gene expression may trigger a faster switch of metabolism between alternative sugars. That faster switch reduces the variance in growth rate by speeding metabolic transitions when conditions change.

Mechanism provides new predictions. Suppose, for example, that increased resources favor high growth rate at the expense of reduced yield. If transporter duplications alter substrate uptake, then enhanced growth may be mediated by the gain of duplicated transporter genes.

Alternatively, suppose that fluctuating conditions favor mechanisms to reduce the variance in growth at the expense of lower average growth. Modified enhancers may reduce grow rate variance by speeding the switch between alternative food sources. The costs for mechanisms of fast switching may reduce overall average growth rate.

In this book, I present many abstract, mechanism-free predictions. Those abstract predictions are simple, general, and broadly applicable.

I also develop many predictions that depend on mechanism. Those mechanism-based predictions provide essential insight into the design of traits. We need both abstract and mechanistic perspectives to enhance our understanding of design.

6.2   Modification of Traits

Comparative predictions forecast the direction of change in traits. Often, we focus on quantitative changes. Grow faster. Secrete less.

When the change is small, heritable variation typically exists or arises de novo. Natural selection can often make small quantitative adjustments in traits.

Large environmental shifts create strong forces, which may favor significantly changed traits. Big changes in traits may depend on enhanced generative processes to provide new sources of variation,

large environmental shift → strong forces → generative processes.

This prediction considers generative processes, including genetic mutation and genomic rearrangement, as traits shaped by the forces of design. A generative process functions by modifying the evolutionary rate of other traits.[308,311]

6.3   Origin of Traits

Upon exposure to a novel toxin, resistance may require a novel mechanism. By what evolutionary sequence does a new resistance trait arise? In general, how do new traits evolve?

GENES FIRST

Perhaps novel genotypes arise by chance. A new genotype may create a new trait or qualitatively alter an existing trait. For example, a new genotype may produce a novel antitoxin or significantly alter an existing antitoxin. First, the genotype arises by chance. Then selection of the genetic variant follows.

The range of traits produced by genetic variants depends on the physical basis by which phenotypic variants arise. For example, if a novel antitoxin requires only a change in the external binding site that attaches to the toxin, then such novelty may arise relatively easily.

By contrast, if existing antitoxins lack the required mechanisms to neutralize a newly encountered toxin, then simply modifying the binding properties of existing antitoxins is not sufficient. Both novel binding and neutralization aspects may be required. Such novelty may rarely arise by just a few simple genetic changes.

The genes-first pathway to novelty has been widely discussed in evolutionary theory.[305] The remainder of this section focuses on an alternative pathway to novelty that has received less attention.

PHENOTYPES FIRST

Perhaps a novel phenotype first appears by variant trait expression among individuals with a shared genetic basis for the trait. Eventually, new genetic variants may heritably stabilize the favored phenotype.[23,132,142,183,268,270,358,431,432,443,451]

For example, cells may use generic pumps to excrete toxins from the cell. Toxin pumping may vary stochastically between cells because pumps depend on a small number of intracellular molecules. Upon initial challenge by a toxin, the survivors may be those phenotypic variants that, by chance, highly express toxin pumps.

Among the survivors, subsequently arising genetic variants may upregulate toxin pump expression, modifying the original trait. Increase of those new genetic variants permanently raises trait expression, stabilizing the favored change.

Cell division rate provides an alternative mechanistic pathway to increased resistance. Suppose the toxin works only against actively dividing cells. Cells vary stochastically in the time between cell division. Quiescent cells resist attack.

Among quiescent cells that survive, a descendant lineage may eventually gain a mutation for a novel resistance trait, such as a modified antitoxin or a variant cell-surface receptor. Increase of the new genetic variant stabilizes the favored change.

In general, a phenotype-first process to generate variability can greatly increase the rate at which traits evolve in response to strong environmental challenge. The evolutionary response may modify an existing trait or create a novel trait.

STOCHASTICITY SMOOTHS THE FITNESS LANDSCAPE

Phenotype-first variation accelerates evolutionary discovery by smoothing the fitness landscape.[132] Modification of an existing trait illustrates the theory. The same principles apply to the origin of new traits.

Suppose that each genotype produces an average trait value, μ. The value of μ varies between genotypes. We can write the probability distribution of phenotypic expression for a given genotype as $p(x|\mu)$, the probability of observing a phenotypic value of x for a genotype with mean value μ. In the following examples, I assume a normal distribution with variance y^2 for all genotypes.

The top row of Fig. 6.1 shows the distribution of phenotypes for a genotype with mean value μ. The solid curve traces a distribution with a relatively small variance. The dashed curve follows a distribution with a relatively large variance.

The second row in that figure shows the fitness, $f(x)$, associated with each phenotype, x. On the left, fitness is high only when the phenotype is very close to the optimum. Other phenotypic values have zero fitness.

The third row shows the average fitness, $F(\mu)$, of a genotype with mean phenotype, μ. The average fitness weights each fitness value, $f(x)$, by the probability, $p(x|\mu)$, of expressing the phenotypic value, x, as

$$F(\mu) = \int_x p(x|\mu)f(x)\mathrm{d}x. \tag{6.1}$$

This transformation for fitness begins with the initial fitness landscape that associates a phenotype, x, with a fitness value, $f(x)$. The distribution of phenotypes, $p(x|\mu)$, expressed by each genotypic value, μ, acts as a smoothing filter to produce the final fitness landscape, $F(\mu)$.

The transformed fitness landscape in the lower left of Fig. 6.1 illustrates the smoothing process. The original fitness landscape, $f(x)$, in the

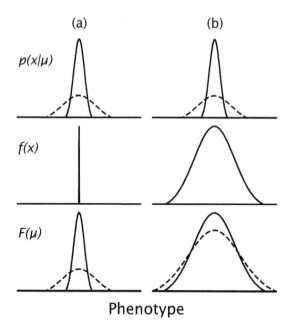

Figure 6.1   Variable phenotypes and fitness. Each column shows how the distribution of phenotypes expressed by a given genotype, $p(x|\mu)$, smooths the fitness function, $f(x)$, to give the expected fitness, $F(\mu)$, for a genotype with average phenotype μ. The smoothing follows eqn 6.1. These examples use normal distributions, $\mathcal{N}(\mu, \delta^2)$. The distribution p has variance $\delta^2 = \gamma^2$, the shape of f follows a curve with variance σ^2, and F follows a curve with variance $\gamma^2 + \sigma^2$ (see Frank[132]). (a) The solid and dashed curves show the phenotypic expression, $p(x|\mu)$, which follows $\mathcal{N}(\mu, 1/2)$ and $\mathcal{N}(\mu, 5)$, respectively. Fitness, $f(x)$, has the shape of a normal distribution with vanishingly small variance, $\mathcal{N}(0, \sigma^2) \to 0$. Thus, expected fitness, $F(\mu)$, is the same as the phenotypic expression, p. (b) The same structure as in (a), except that $f(x)$ is much wider, following $\mathcal{N}(0, 7)$. Thus, $F(\mu)$ now follows $\mathcal{N}(0, 7.5)$ and $\mathcal{N}(0, 12)$ for solid and dashed curves, respectively. In each plot, the baseline is set to 4.3% of the peak in that plot. The baseline truncates phenotypes with low vigor, setting their fitnesses to zero. From Figure 2 of Frank.[132]

panel above is a narrow peaked function. To survive and obtain nonzero fitness, a phenotype must almost exactly match a specific expression.

If each genotype produces a particular phenotype without any variation, then, using Maynard Smith's[268] language, matching a genotype to the favored form would be like searching for a needle in a haystack. By contrast, if each genotype produces a distribution of phenotypes, then, as in the bottom row, matching a genotype to the favored form would be

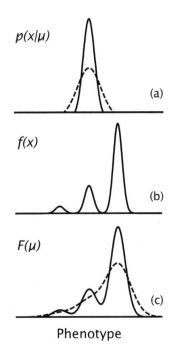

Phenotype

Figure 6.2   A broad expression of phenotypes smooths a multipeak fitness landscape. (a) The dashed curve shows broader phenotypic expression, $p(x|\mu)$. (b) The fitness landscape, $f(x)$, has multiple peaks. (c) Broad phenotypic expression (dashed curve) smooths the realized fitness landscape, $F(\mu)$, to a single peak. In this example, the narrow and broad phenotypic expression patterns follow $\mathcal{N}(0, y^2)$ with variances of 0.04 and 0.16, respectively. Fitness is given by $f(x) = \int_{y=-1}^{1}[3(1+y)^2 + 1]\mathcal{N}(y, \sigma^2)dy$, with $\sigma^2 = 0.0225$. The value of $F(\mu)$ is calculated from eqn 6.1, yielding the expression for $f(x)$ in the prior sentence with the variance replaced by $\sigma^2 + y^2$. The baseline truncates small values. From Figure 4 of Frank.[132]

like searching for a needle in a haystack when someone tells you when you are getting close.

Put another way, phenotypic smoothing transforms the problem of exactly matching a target phenotype, without any clue about how close the current genotype is, into the problem of climbing a smooth fitness gradient to a local peak. Natural selection is very bad at finding special traits for which nearby phenotypes have low fitness. Natural selection is very good at improving traits by climbing a fitness gradient toward a local peak.

PEAK SHIFTS ON SMOOTHED LANDSCAPES

Multipeak fitness landscapes provide the classic model for the origin of new traits.[62,455,456] Figure 6.2b illustrates a multipeak landscape. Suppose the phenotypes of a population cluster near the middle peak. That population cannot evolve toward the higher peak by climbing a fitness gradient.

Put another way, small quantitative modulations of the current trait cannot improve performance. Instead, a qualitatively distinctive shift in the trait may be required to achieve the higher peak.

How can a population discover the improved trait when small modifications of the existing trait reduce fitness? Stochasticity of phenotypic expression provides one solution. Stochasticity can smooth the fitness landscape, transforming the difficult peak-shift problem into the simple problem of climbing a smooth fitness gradient.[132]

Suppose the phenotypic expression for a given genotype varies. Figure 6.2a shows two examples for stochastically variable expression, $p(x|\mu)$, which is the distribution of phenotypes, x, for a genotype with mean phenotype, μ. Once again, we can apply eqn 6.1 to obtain the fitness landscape that matches each genotype's mean phenotype, μ, to its fitness, $F(\mu)$.

Small amounts of phenotypic stochasticity, shown by the solid curves, partially smooth the landscape in Figure 6.2c. However, distinctive peaks remain. The population cannot climb a smooth gradient from the middle peak to the higher peak.

Greater phenotypic stochasticity (dashed curve) smooths the landscape into a continuously rising gradient toward a single peak. Natural selection can push the population up the smooth gradient to the peak.

Increased stochasticity transforms the difficult problem of shifting between peaks into the easy problem of climbing a hill. Phenotypic stochasticity accelerates the discovery of novel trait expression.

PHENOTYPIC PLASTICITY AND NEW TRAITS

Stochasticity accelerates evolutionary rate. However, stochasticity typically confines phenotypes within an existing set of trait values.

How can evolution discover qualitatively distinctive traits in response to environmental challenge? New genetic variants may produce novel traits. However, creating complex novel traits in one step may be difficult.

Instead, novel phenotypes may first arise when individuals respond to new environments by adjusting their development, physiology, or behavior. This phenotypes-first pathway for generating novelty may play an important role in the evolution of new traits.[443]

If novel traits created by phenotypic plasticity partially meet the new environmental challenge, then natural selection of genetic variants can favor quantitative improvements of those novel traits, climbing a local fitness gradient. This process transforms the difficult problem by which evolutionary process discovers novelty into the simple process by which natural selection climbs a hill to improve an existing trait.

Comparatively, greater phenotypic plasticity increases the rate at which novel traits evolve in response to environmental challenge. For example, the rate at which a microbe acquires resistance to a novel toxin or drug may increase with the microbe's plastic response to environmental stress. Consider a speculative scenario to illustrate the process.

Suppose a microbe comes under attack by a novel toxin. The toxin enters the cell by binding to a cell-surface receptor. If the cell can survive for a period without the receptor, then shutting down receptor expression in response to the toxin may allow the cell to resist attack.

To shut down receptor expression, the cell must evolve a sensor for the toxin and the regulatory pathways that link the sensor to the expression of the receptor. Simultaneous evolution of coordinated components is difficult. The sensor alone provides no advantage. A regulatory switch provides no advantage without appropriate sensors.

Genetic mutations rarely provide such simultaneous jumps in multiple traits. A genes-first pathway for the origin of complex traits seems difficult. Put another way, no simple gradient of increasing fitness exists for natural selection to climb.

However, cells often have pre-existing regulatory pathways for responding to environmental cues. Such phenotypic plasticity greatly enhances the opportunity to evolve new traits. For example, a generic cellular response to attack may shut down the expression of several receptors, including partial reduction of the target receptor for the toxin.

The existing plasticity of the cell already creates a phenotype that is close to the required trait. In the next evolutionary step, a novel mutation may link an existing sensor, partially stimulated by the toxin, to the general defense response. That step would climb up the fitness gradient, in which a direct change in one character raises fitness.

Once a specific receptor is linked to the regulatory control that switches expression, further genetic variants that modulate the response may be favored by climbing the hill of increasing fitness.

In this scenario, the evolutionary path started with a phenotypic variant induced by the cell's intrinsic phenotypic plasticity. Then genetic variants in sensors and regulatory wiring improve the initial phenotype. The difficult problem of novel trait discovery transforms into the simple problem of continual improvement by small changes.

PHENOTYPES-FIRST PATHWAY ACCELERATES EVOLUTION

A genotypes-first pathway for new traits is easy to understand. Mutations arise randomly. Large populations contain many mutants. A new environmental challenge favors one of the pre-existing mutant genotypes.

The genotypes-first pathway suffers from a major difficulty. Selection can favor only those traits that originate by genetic changes, which we may broadly call *mutations*. Discovering novel traits by mutation may not happen easily. For example, how do two mutually beneficial traits arise when neither one alone provides value?

A phenotypes-first pathway can accelerate trait discovery. Each existing genotype produces a range of phenotypes. The first step requires only that a genotype express a trait that gains a little bit with respect to a novel challenge. That first favored form arises by nonheritable phenotypic variability.

Subsequently, selection favors those genotypes that can produce phenotypes more closely matched to the target. The difficult problem of discovery transforms into the relatively simple problem of hill climbing.

Phenotypes first does not automatically discover two synergistic traits. But it makes discovery easier because plasticity often responds to environmental challenge by modulating suites of interacting mechanisms.

By covarying mechanistic components, plasticity generates phenotypic variety that may be in the right direction with respect to a novel challenge.[443] That phenotypes-first step toward discovery smooths the fitness landscape. A smoothed landscape provides a more direct path of genotype change, accelerating the process of novel trait evolution.

Despite the long history and extensive literature on phenotypes-first trait discovery, this aspect of adaptation remains an underappreciated evolutionary force. Comparatively, greater stochasticity of trait expression or greater phenotypic plasticity increases the rate of novel trait discovery, mediated by the smoothing of the fitness landscape.

7 Theory: Control

How do organisms adjust their traits to changing conditions? When perturbed, how do traits return to their target values? Responsiveness and homeostasis pose universal challenges of regulatory control.

The key comparative question is: How do different environments change the favored design of regulatory control? To address that question, this chapter introduces the various challenges that control systems must solve. Those multiple challenges lead to key tradeoffs in the design of regulatory control.

Consider, for example, a receptor that takes up a food source. Ideally, more available food stimulates greater receptor expression. Less food reduces expression. How does the cell adjust?

Food could directly stimulate receptor production. If receptors decay at a constant rate, then raising production in proportion to food availability is sufficient to control receptor number.

In that feedforward control process, information flows in one direction. The external signal of food availability alters the internal process of production. The production rate in response to food availability can be designed to give a particular target level for receptor number, as long as the decay rate of receptors remains constant.

Feedforward control cannot correct errors. If some unknown process destroys receptors more quickly than normal, the actual number of receptors would not match the ideal number. Feedforward control has no process to reduce errors.

The first section introduces error-correcting feedback. The difference between the target value and the actual value feeds back as input into the control process. With a measure of error, the system improves simply by adjusting to reduce the error. Error-correcting feedback is perhaps the greatest principle of design.

The second section contrasts feedforward and feedback control. The example describes homeostatic maintenance of a setpoint in response to environmental perturbations.

The third section notes that error-correcting feedback typically requires signal amplification. A system reduces error more quickly when it amplifies the signal carrying information about the error. Signal amplification may require additional energy, a cost of error correction. Signal amplification may also cause instability, another cost of error correction.

The fourth section develops the benefits of error-correcting feedback. When system dynamics are uncertain, feedforward cannot correct mismatches between the target and the actual output. By contrast, error-correcting feedback robustly adjusts for uncertainties.

Error correction also compensates for poorly performing internal system components. That reduced pressure on component performance often favors designs with less costly, lower-performing components. Error-correcting feedback at the system level begets more errors at the component level, the paradox of robustness.

The fifth section develops the tradeoff between responsiveness and homeostasis. Better responsiveness requires rapid adjustment to environmental change. However, the more rapidly a system adjusts to change, the more sensitive the system becomes to disturbances that perturb a homeostatic setpoint.

The sixth section raises problems of sensor design. Sensors obtain information about the external environment and the internal state. How should sensors be tuned for their sensitivities to different frequencies of change? How can arrays of sensors be designed to improve control? How do different environments favor distinct sensor designs?

The seventh section summarizes various tradeoffs in control design. The tradeoffs lead to comparative predictions about control.

7.1   Error-Correcting Feedback and Robustness

Authors often note the precision of molecular control and response.[198,375] Yet, at a small scale, biology is anything but precise. It is instead highly stochastic and error-prone.[94,199]

Some molecules occur in low, widely fluctuating numbers. Changes in temperature affect fluctuating molecular motion, binding rates, and reaction times. Environmental signals mix useful information with noise. Sensors are imperfect. Mutation and recombination alter components and swap parts.

How does precise function arise from such a mess? Finely tuned components are, at best, costly to produce and maintain, and often cannot be made. How do biological systems design robust performance in spite of sloppy components?

Error-correcting feedback is the great principle of robust design. The error measures the difference between a system's actual output and its target. By feeding back the error as an input, the system can move in the direction that reduces the error.[70,89,138,471]

Error correction compensates robustly for misinformation about system dynamics and for perturbations to system components. Excellent performance often follows in spite of limited information, sloppy components, and noisy signals.

The following sections emphasize general principles. Those principles provide the foundation on which to develop specific predictions.

7.2   Principles of Control

Control transforms environmental and internal inputs into biological outputs. Resource gradients stimulate motion. High sugar concentration increases matching cell surface receptors. Slow temperature fluctuations cause metabolic adjustments. Rapid temperature fluctuations are homeostatically buffered, leaving metabolic processes unchanged.

What regulatory control design best modulates those input-output transformations? Comparatively, how do we expect a changed pattern of inputs to alter regulatory design? For example, as input sugar concentration fluctuates more rapidly, how should regulatory design change to cope with the increased noise in the input signal?

The following summary highlights a few key aspects from earlier publications.[138-140]

WHAT IS DESIGN?

Design implies a strong statement about how something came to be.[446] Molecular systems and mechanisms of regulatory control can be particularly difficult to interpret. If we see the production rate of a molecule increase with temperature, is that a designed feature?

Reaction rates tend to increase with temperature for physical reasons. Often, an increasing reaction rate with temperature must be regarded as an inevitable physical consequence rather than a designed feature. But some increases in reaction rate seem to be designed responses to enhance performance, such as a rise in the production rate of heat shock proteins.[244] In general, a modifiable component that has been tuned to achieve some goal forms part of a designed system.

Consider a biochemical system that includes both production and degradation rates of some molecule. If the regulatory control system modifies the degradation rate to track some target setpoint of molecular abundance, then degradation acts as a designed error-correcting feedback mechanism. In this case, degradation rises and falls with the error between the current molecular abundance and the target abundance.

WHAT CONDITIONS FAVOR ERROR-CORRECTING FEEDBACK?

> There are two, and only two, reasons for using feedback. The first is to reduce the effect of any unmeasured disturbances acting on the system. The second is to reduce the effect of any uncertainty about systems dynamics.
>
> —Glenn Vinnicombe[426]

A system transforms inputs to outputs. The system design must handle unpredictable external disturbances and unpredictable internal components that alter the dynamics of the transformation.

An error-correcting design robustly corrects for uncertainties. But error correction requires costly additional machinery. When do the benefits outweigh the costs? Comparatively, how do changed conditions alter the predicted tendency for error-correcting designs?

This subsection focuses on uncertainty about system dynamics. How does error correction compensate for unpredictable internal components? How much benefit does an error-correcting design provide when compared to a design without error correction?

The theory concerns biological function, as opposed to the biophysical details of particular mechanisms. The functional principles guide the development of comparative predictions.

Begin with a fixed process, P. The process takes input u and produces output y, as in Fig. 7.1a.

System design modulates the input-output transformation to achieve particular goals. Contrast two alternative designs. In Fig. 7.1b, a second process, C, controls the input into P. In biology, the fundamental forces of design shape C. For simplicity, we can think of this problem as an

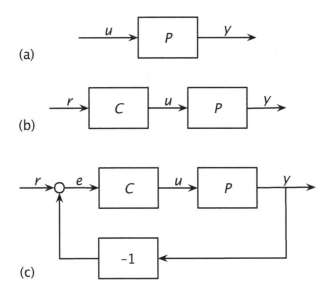

Figure 7.1 Alternative designs for control of a process, P. (a) The intrinsic uncontrolled process, taking input, u, and producing output, y. (b) Modulation of the input signal, u, by a designed control process, C. The controller takes an external input, r, which may be an environmental variable or an internal reference signal that defines the system's target setpoint. In this feedforward open loop, inputs flow to outputs without feedback. (c) A closed loop feeds back the error, $e = r - y$, as the system input. Redrawn from Frank.[138]

engineering task in which we design C optimally, subject to given target goals, tradeoffs, and constraints.

The controller, C, takes an input, r. The input may come from an environmental sensor, providing information about sugar concentration, temperature, or other external or internal environmental attributes. Or the input may come from another internal component that sets the desired output of our system.

The system in Fig. 7.1b does not have error-correcting feedback. The feedforward open loop follows a direct and continuous path of transformations from the external signal, r, to the control signal, u, and then to the final output, y.

Figure 7.1c shows error-correcting feedback. The output, y, feeds back to the input through a closed loop. The error, $e = r - y$, is the difference between the current input and output. If r is the target setpoint for the system, then we design the system to increase its output when the error is positive and decrease its output when the error is negative.

To track the setpoint input, r, in the open loop of Fig. 7.1b, we must know the process, P, to design the controller, C. Any misinformation about P or unknown disturbance of the signals leads to a mismatch between the actual output and the target output.

In the closed error-correcting loop of Fig. 7.1c, the system receives continuous updating of its distance and direction from its target. The system can correct for misinformation about the dynamics of P and for any unknown disturbances to the system.

EXAMPLE OF PROCESS AND SYSTEM DESIGN

The benefit of error correction depends on the kinds of disturbances and uncertainties. This subsection illustrates uncertainty in process dynamics. See Frank[139] for details.

Process dynamics.—The alternative system designs in Fig. 7.1 show process dynamics as P, an unspecified process that takes input u and produces output y. To develop a specific example, assume that P follows a basic second-order differential equation,

$$\ddot{x} + \alpha\dot{x} + x = u. \tag{7.1}$$

For $a_1 + a_2 = \alpha \geq 2$ and $a_1 a_2 = 1$, we can write the system as a pair of first-order equations,

$$\dot{x}_1 = -a_1 x_1 + u \tag{7.2a}$$
$$\dot{x}_2 = -a_2 x_2 + x_1, \tag{7.2b}$$

with system output $y = x_2$. Figure 7.2a shows the mechanistic interpretation of this system. In the first step, an external input, u, drives production of x_1, which has a constant decay rate of a_1. This mechanism is among the simplest and most common processes, in which stimulated production is balanced by intrinsic decay. The dynamics describes an exponential process in which the level of x_1 at time t is

$$x_1(t) = \int_0^t e^{-a_1\tau} u(t - \tau)d\tau$$

when starting from an initial value of zero. For constant input, u, the value is

$$x_1(t) = \frac{u}{a}(1 - e^{-a_1 t}),$$

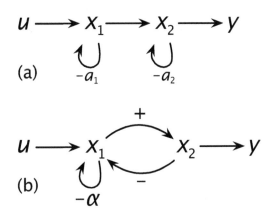

Figure 7.2   Examples of system process, P. These examples show alternative
mechanisms for second-order dynamics. (a) A cascade of exponential processes.
The incoming signal u stimulates production of x_1, which degrades at rate
a_1. The level of x_1 stimulates production of x_2, which degrades at rate a_2.
Dynamics given in eqn 7.2. (b) The first part of this mechanism is the same as
the upper panel, with u stimulating production of x_1, which degrades at rate α.
In addition, x_1 and x_2 are coupled in a negative feedback loop. Dynamics given
in eqn 7.3. Redrawn from Frank.[139]

which, over time, approaches the ratio of the stimulation rate divided by
the degradation rate, u/a. The second step in the mechanistic cascade
of Fig. 7.2a is also an exponential process, with input x_1 and output
$y = x_2$.

When $\alpha < 2$, the mechanistic basis cannot be split into a cascade of
separate exponential processes. For any real value of α, including $\alpha \geq 2$,
we can rewrite the second-order system in eqn 7.1 as a different pair of
first-order processes,

$$\dot{x}_1 = -\alpha x_1 - x_2 + u \qquad (7.3a)$$

$$\dot{x}_2 = x_1, \qquad (7.3b)$$

with system output $y = x_2$. Figure 7.2b shows this process as a negative
feedback loop between two components, x_1 and x_2. Negative values of
x may arise. We can add constants to prevent negative values.

When $\alpha = u = 0$, the system is a pure oscillator that follows a sine
wave. For $0 < \alpha < 2$ and $u = 0$, the system follows damped oscillations
toward the equilibrium at zero because the degradation of x_1 at rate

$-\alpha$ causes a steady decline in the amplitude of the oscillations about the equilibrium. Overall, eqn 7.1 describes several basic mechanistic processes and associated dynamics.

Performance metrics.—The squared distance between the system's target setpoint and its actual output provides a common measure of performance. The sum of the squared deviations over time measures the total performance.[13,138,303,471]

We must consider two major aspects of performance. First, how well does the system respond to an environmental input that changes for a significant period of time? Second, how well does the system reject brief environmental fluctuations and maintain its internal homeostasis?

Control theory evaluates responsiveness by analyzing how closely a system tracks a step change in the input setpoint. In particular, suppose the system initially adjusts to a constant input, $r = 0$. Then the input is changed in a step to $r = 1$ and kept at that level. How closely does the system match the input over a period of time?

Suppose the step change happens at $t = 0$. Then we may measure the total performance over T units of time in response to a step change as

$$J_s = \int_0^T (e^2 + \rho\tilde{u}^2)\,\mathrm{d}t. \tag{7.4}$$

Here, the error, $e = r - y$, measures the deviation from the setpoint, in which each variable is a function of time. Thus, e^2 is the squared distance from the setpoint.

The second term, $\tilde{u} = u - r$, captures the cost of the control signal. In Fig. 7.1, the designed systems modify the external signal, r, with a control process, C, that sends the control signal, u, to the fixed system process, P. Modifying the external signal by the control process, C, may require energy or other costs. Thus, the amount of change in the signal, $\tilde{u}$, may associate with the cost of control.

Homeostatic maintenance of system output can be measured in various ways. Control theory typically analyzes the system's response to a large instantaneous perturbation. Technically, $r = 0$ at all times, except at a single instant when r becomes infinitely large.

If we apply the perturbation at time zero and measure the system's subsequent deviation from the zero setpoint, then we measure the response to a perturbation as

$$\mathcal{J}_p = \int_0^T (y^2 + \rho u^2)\,dt. \tag{7.5}$$

Because $r = 0$, the squared error $e^2 = (r - y)^2$ reduces to y^2, the square of the system output. Similarly, $\tilde{u}^2 = (r - u)^2 = u^2$.

For both measures, smaller values of $\mathcal{J}$ mean smaller distances from the optimal trajectory. Thus, lower values of $\mathcal{J}$ associate with greater performance.

We may combine these measures of responsiveness and homeostasis into an overall performance metric as

$$\mathcal{J} = \mathcal{J}_s + y\mathcal{J}_p, \tag{7.6}$$

in which y describes the weighting of the homeostatic performance in response to perturbation relative to the tracking performance in response to a step change in the environmental reference signal. Optimal performance minimizes $\mathcal{J}$.

Optimal control.—First, we consider the intrinsic process by itself. The system, shown in Fig. 7.1a, is $u \to P \to y$. The process, P, given by eqn 7.1 and repeated here for convenience,

$$\ddot{x} + \alpha\dot{x} + x = u,$$

depends on the single parameter, α.

The value $\alpha = \sqrt{1 + y}$ optimizes the performance metric $\mathcal{J}$ (eqn 7.6).[139] A minimal value of $\mathcal{J}$ means that the system tracks as closely as possible to the target value. Here, we assume no cost for amplifying control signals, thus $\rho = 0$ in the calculations of $\mathcal{J}_s$ and $\mathcal{J}_p$.

The optimized system remains second order because we have only the single parameter α to tune. This optimized second-order system remains prone to oscillation and to overshooting target output values.

Adding a controller process to modulate the input, u, can improve performance. We have two different control designs to consider, the open loop in Fig. 7.1b and the error-correcting closed loop in Fig. 7.1c.

Both architectures for control take an input, r, and produce an output, y. We can think of both systems as $r \to G \to y$, in which G describes all of the dynamics and transformations of the input that ultimately produce the output. The characteristics of the internal processing, G, differ between the two architectures.

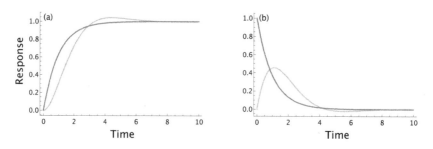

Figure 7.3   Dynamics of the intrinsic process, P, from eqn 7.1, with $y = 1$ and optimal parameter $\alpha = \sqrt{1+y} = \sqrt{2}$ (light curves) compared with the optimized open and closed loop systems with process G in eqn 7.7 with $p = 1/\sqrt{y} = 1$ (dark curves). (a) The unit step response. (b) The impulse perturbation response. From Frank.[139]

Suppose we take the dynamics of P as known and fixed at the optimum with $\alpha = \sqrt{1+y}$, and find the optimal control process, C. That optimal control process differs between the two architectures because of the distinct ways in which open and closed loops transform input signals.[138,471]

However, the best overall internal processing, G, has the same dynamics for both architectures when the control process is limited to a second-order differential equation in its input, r, and output, u. In particular, G is given by[139]

$$\dot{x} = -p(x - r), \tag{7.7}$$

with $p = 1/\sqrt{y}$ and final output, $y = x$. Figure 7.3 compares the dynamics of the optimized process, P, without additional modulation by control (light curves) and with modulating control (dark curves).

The uncontrolled process has reasonably good response characteristics because we chose the parameter α to optimize performance. For $y = 1$, the uncontrolled process has performance $\mathcal{J} = \sqrt{2}$. Optimized control improves the response characteristics, yielding an improved performance metric, $\mathcal{J} = 1$.

7.3   Error Correction and Signal Amplification

The optimized open and closed loops have the same overall system dynamics in eqn 7.7. However, they construct those dynamics in different ways. The following example illustrates the differences.

Steering a Car

Suppose you are the driver, acting as the system's controller, C. You produce the control signal, u, to alter the control of the car's internal steering process, P, which sets the car's direction, y. In other words, you move the steering wheel, which changes the car's mechanism for setting the direction.

In an open loop, you receive the input signal, r, which is the current direction of the road. In this case, you cannot see the actual direction of the car, as if you were remotely steering a car you cannot see based on the current direction of the road, which you can see.

You move the steering wheel to match the input signal's direction. You believe the steering mechanism is accurate, keeping the unseen direction of the car on its course. But you cannot check. An inaccurate steering mechanism misaligns the car. You cannot fix that error.

As the input signal about the road's direction changes, you apply mild pressure to the steering mechanism to adjust its setting for the desired direction. Because you do not know the car's current direction, you do not know how close or far off you are from matching the car's direction to the road.

The match depends on how accurately you perceive the road's direction, how accurately you transform the perceived direction into steering changes, and how accurately the steering mechanism adjusts the car's direction. The car remains on course only when each step is accurate.

Steering in an error-correcting feedback loop is different. You perceive the input signal, $e = r - y$, which is the error between the road's direction and the car's direction. With information about the error, you steer by adjusting to the error. When the car is too far left, you turn right. When too far right, you turn left.

Signal Amplification

The greater the error, the harder you turn. The harder you turn, the faster you correct the error. Put another way, greater amplification of the error signal more rapidly reduces the error. Or, as stated by Åström & Murray [17, p. 320]

> Feedback and feedforward have different properties. Feed-
> forward action is obtained by ... precise knowledge of the

process dynamics, while feedback attempts to make the error
small by dividing it by a large quantity.

Dividing the error by a large quantity is equivalent to amplifying the
error signal to reduce the error more rapidly. The harder you turn to
correct steering errors, the more quickly you reduce the error.

Comparatively, error-correcting systems tend to amplify control signals more strongly than do open loop feedforward systems. The faster a
system must reduce errors, the more strongly it tends to amplify error
signals. Error correction and signal amplification are among the great
principles of systems design.

STABILITY VERSUS PERFORMANCE

Signal amplification causes strong responses, which can make a system
prone to instability.

Consider steering. If you are to the left of your target direction and
turn hard to the right to compensate, the car may swerve past the target
and end up too far to the right. Turn too hard back to the left and you
may overshoot again. Overshooting risks instability and total loss of
control. Unstable demise is a major risk of error correction with high
signal amplification.

The more strongly instability poses a systemic risk, the greater the
need for a design to include a margin of safety. Typically, a broader
stability margin requires reducing signal amplification and slowing the
response to errors. A slower response lowers performance.

7.4   Robustness to Process Uncertainty

In the car example, error correction adjusts for inaccuracies in the
steering mechanism. If you are off course because of inaccurate steering,
the observed error tells you what you need to do to get back on course.

In general, error correction compensates for variability in process
dynamics. Figure 7.4 illustrates error-correcting robustness to process
variability for the second-order dynamics of eqn 7.1. In that equation,
the parameter, α, determines process dynamics.

As α varies from its optimal value, the performance cost metric, J,
increases. Larger values of J associate with greater distance from optimal
tracking and thus with worse performance.

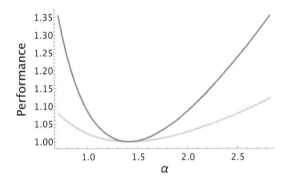

Figure 7.4   Sensitivity of open loop (dark curve) and closed loop (light curve) systems to parametric variations in dynamics. The y-axis shows the performance cost metric, J, and the x-axis shows the process dynamics parameter, α. The optimal value is at $\alpha = \sqrt{2}$, which minimizes the performance, J. From Frank.[139]

The figure shows the great sensitivity and lost performance for the open loop without error correction (dark curve). Performance degrades rapidly as α varies from its optimal value. Open loop control cannot compensate for the misspecification of process dynamics.

The performance of the error-correcting process degrades relatively little with variation in process dynamics (light curve). In that case, the constant input of the error between the target output and the actual output allows the system to improve its performance without prior information about system dynamics.

Robust design in life and in human engineering would not be possible without error correction. Building and maintaining precise components are very costly and perhaps impossible. With error correction, systems can perform robustly with sloppy components and limited information.

THE PARADOX OF ROBUSTNESS

> The ultimate effect of shielding men from the effects of folly,
> is to fill the world with fools.
>
> —Herbert Spencer

The better a system becomes at compensating for perturbations and errors, the better the system can handle imprecise and erratic system components. System robustness weakens the selective pressure acting on the system's components. Weaker selective pressure on the components leads to decay in their performance.

Overall, the more robust a system, the more the system's components will tend to decay in performance. Better error correction begets more errors.[124,125,134]

Component decay may take the form of increased variability or sloppiness in function. Alternatively, weaker selective pressure on components may cause them to decay to less costly and lower performing designs. In the latter case, the economics of efficiency favors robust systems to use cheaper components.

Comparatively, the better a system is at buffering fluctuations in its components, the more the components will tend to accrue genetic variability[349,427,435] and stochastic variability in expression, and the more those components will tend to decay to cheaper, lower performing states.[125]

Within systems, performance will be more sensitive to some components than to others. The less sensitive the system is to fluctuations in a particular component, the more variable and lower performing that component will tend to be.[140]

In theory, the paradox of robustness should be a ubiquitous aspect of control systems. If so, the paradox likely plays a key role in biological design throughout the history of life. However, this aspect of design has received little attention in theory or application.

7.5   Responsiveness versus Homeostasis

Microbes must respond to environmental change. When a new food source arrives, cells change to acquire and digest the food. Faster response of the required traits typically provides a benefit.

Microbes must also maintain steady expression levels in a stable environment. Homeostatic maintenance poses a challenge because environmental signals often fluctuate significantly over short time periods. Slower response to changing signals typically improves homeostasis.

Overall, fast response improves tracking of true environmental change, whereas slow response improves homeostatic maintenance relative to noisy environmental fluctuations. This tradeoff between responsiveness and homeostasis shapes many aspects of design.[89,138,303,471] Figure 7.5 illustrates the tradeoff.

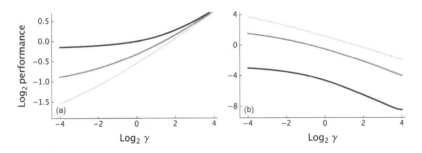

Figure 7.5   The tradeoff between responsiveness and homeostasis. Smaller performance values correspond to better performance. The parameter y from eqn 7.6 determines the relative weighting between responsiveness to an environmental change and the ability to maintain a homeostatic setpoint when perturbed by a single large impulse disturbance. Larger y weights homeostatic performance more heavily. The different curves in each panel show different costs, ρ, for the control signal, u. The control signal penalty, given as $\log_{10} \rho$, is -4 (light line), -2 (medium line), and 0 (dark line). (a) Responsive performance to a step increase in the environmental input signal, measured by J_s in eqn 7.4. (b) Homeostatic performance in response to an impulse perturbation in the environmental input signal, measured by J_p in eqn 7.5. Redrawn from Frank,[139] which also provides details on the assumptions and the optimization methods used to obtain the curves.

All curves arise from minimizing the performance metric in eqn 7.6, repeated here for convenience,

$$J = J_s + y J_p.$$

The first component, J_s, is the total distance between the system's output value and the recently shifted target value caused by a long-term environmental change. Smaller values mean closer tracking of the environment and more successful performance.

The second component, J_p, is the total distance between the system's output value and its homeostatic setpoint. At time zero, the system receives a large instantaneous environmental impulse, which then immediately disappears. The impulse perturbs the system from its setpoint. Greater perturbed deviation and longer time of return to the setpoint increase J_p. Larger performance values associate with worse homeostatic maintenance and less successful performance.

The parameter y determines the relative weighting between the two components of performance. Larger y weights the homeostatic component more heavily.

Figure 7.5a shows the responsiveness component of performance. The performance value rises and success declines as y increases because a stronger weighting of homeostasis reduces responsive success. In essence, the system responds more slowly, protecting homeostasis from being perturbed by fluctuations but simultaneously slowing the system's ability to respond to real change.

Figure 7.5b shows the opposing change in homeostatic performance. Larger y values weight this component more strongly, favoring lower (better) performance values.

The different curves in each panel of the figure correspond to different cost weightings, ρ, for the control signal, u. As ρ rises, control signals become more costly. In this example, responsiveness requires a stronger control signal than does homeostasis. As the cost of the control signal rises, responsiveness becomes relatively expensive, altering the balance of the system toward favoring homeostasis over responsiveness.

The tuning of system design for responsiveness versus homeostasis also depends on the frequency spectrum of environmental change. When environments tend to change slowly, at low frequency, systems gain by being more responsive to altered conditions. When environments fluctuate rapidly, systems gain by improving homeostasis, which typically means that they respond more slowly to change.

Put another way, as systems become better at coping with high-frequency noise, they often degrade in the speed of their response to longer-term changes. Alternatively, as they become better at responding rapidly to change, they become more sensitive to short-term perturbations that disrupt homeostasis.

One can plot a system's response to different frequencies of inputs. Such Bode plots play a central role in engineering control design.[89,138,303,471] How can one tune a system to be more responsive to certain input frequencies and less sensitive to perturbations at other frequencies? And, comparatively, how does a change in responsiveness at particular frequencies alter the predicted mechanistic attributes of trait design?

7.6 Sensors

Matching the environment requires external sensors. Maintaining homeostasis requires internal sensors.

Sensor design raises several challenges. What frequencies of environmental change provide useful information? What frequencies should be ignored? Typically, low-frequency signals represent long-term environmental changes that require adjustment. High-frequency fluctuations represent noise that should be ignored.

Each sensor has a particular response profile to different frequencies. How best to tune frequency response? Mechanistically, how are sensors built from biological components?

Sensors can be thought of as estimators of parameters given some data. Optimal estimation often follows from analysis of Fisher information, a way of quantifying the information in data about a parameter of a probability distribution.[44]

That information abstraction only gives a hint about the many problems of sensor design. Individual sensors necessarily have particular frequency and magnitude sensitivities. Given variable sensor sensitivity, what is the best design of sensor arrays to meet particular environmental challenges?[137,196]

What is the best temporal and spatial deployment of sensors? How does one balance the benefits of information versus the costs of sensors and the use of the acquired information?

Comparatively, as the frequencies of meaningful signals change, how does sensor design change? How does a change in the magnitude of key signals alter sensor design?

Sensors provide a great challenge in the study of biological design. A broad comparative theory has yet to be developed. Only through comparative predictions can one study the forces that shape design.

7.7 Control Tradeoffs

This chapter emphasized tradeoffs in the regulatory control of trait expression.[139,140]

OPEN LOOP VERSUS ERROR-CORRECTING CLOSED LOOP

Error correction provides benefits when a system does not have perfect information about disturbances and dynamics. By measuring the error between the target output and the actual output, a system can continuously adjust itself to improve performance. It can also compensate for sloppy or faulty components.

Error correction requires additional sensors to measure the error and strong signal amplification to reduce the error. Simpler open loop control may be better when a system has good information about its dynamics or the costs of error correction are high.

Comparatively, error correction is more strongly favored as information about dynamics declines, perturbations increase, or components become less reliable. In other words, robust design becomes more strongly favored as dynamics become less predictable.

ROBUSTNESS AND DECAY

The more robust an error-correcting system becomes, the more component performance tends to decay. By correcting for variable or faulty components, error correction reduces the benefit of precise and highly optimized components. As the benefit and associated force favoring component performance declines, the components tend to decay to lower cost and lower performance. Components may also become more variable in their output. Robust system design may often lead to highly stochastic system components.

PERFORMANCE VERSUS STABILITY

High performance often means adjusting quickly to a changed environment. Fast adjustment requires a strong force in the direction of the new target. The stronger the force, the more rapid the adjustment and the greater the chance of overshooting the target. Overshoot requires a strong countering force to reverse direction. Overshoot in the opposite direction may occur.

Increasingly larger overshoots cause instability. Typically, the danger of instability rises with the speed of adjustment. To prevent dangerous and potentially lethal instability, systems may pay the cost of reduced performance in order to lower the risk of instability. The greater the margin of safety is, the lower the average performance will be.

Comparatively, environments that impose greater perturbations favor systems designed with greater stability margins and lower average performance.

RESPONSIVENESS VERSUS HOMEOSTASIS

Systems must respond to long-term environmental change and avoid being perturbed by short-term noise. These goals of responsiveness and homeostasis often trade off against each other. A more responsive system adjusts rapidly to environmental change. But a system that responds quickly is sensitive to rapidly fluctuating false signals.

Comparatively, noisier environments favor greater homeostasis at the expense of responsiveness. By contrast, dominant low-frequency environmental changes favor greater responsiveness. Greater cost of control typically favors homeostasis over responsiveness because a responsive design tends to require costly internal signal amplification to track environmental change.

SENSOR DESIGN

The theory of sensor design requires greater development. Some obvious tradeoffs arise. In sensor arrays, more low quality sensors trade off against fewer high quality sensors. Greater numbers of sensors provide better spatial coverage. Combining inputs from multiple sensors may require more cost to compute the output signal from the multiple inputs.

When tuning sensors, what factors favor particular frequency bands of sensitivity and particular magnitudes of sensitivity? What is the best combination of sensitivities in an array? Mechanistically, how do cells achieve sensor designs with particular functional attributes? Comparatively, how do changes in environmental parameters alter sensor design?

8 Studying Biological Design

[A]ll observation must be for or against some view if it is to be of any service!

—Charles Darwin[76]

What does it mean to say that we understand biological design? We must understand how particular causal forces have shaped the designs that we see in nature.

How do we match a causal force to the shaping of design? We must build on the idea that a change in force causes a change in state.

Why a change in force? Because many forces may initially be in balance. We identify a force and its consequences when the force changes.

How do we infer forces when studying biological design? Usually, we do not observe forces directly. Instead, we must infer what we cannot see. We suppose that an unseen force mediates between an observable partial cause and an observable effect, $P \to F \to E$.

What is an observable partial cause? A change in something measurable that we think may have an effect. Greater resource flow. Increased lifespan of resource patches. More competition between genotypes.

Why the *partial* qualifier for a cause? Because every effect has many causes. Each cause partially determines outcome. Many things alter growth rate. Increased resource competition may be one partial cause.

What is an observable effect? A change in any measurable attribute that we consider related to design. A trait, a tradeoff, the amount of variability in something, the dynamic tendency to fluctuate, the way an organism adjusts to its environment.

What is a mediating force? A process that mediates between cause and effect, defined inductively by the accumulation of reasonable interpretations. Kin selection mediates between a causal change in the genetic similarity of competitors and the effect on growth rate.

Why is *cause* used for both the observable change in the condition, P, and the mediating force, F? The observable change that drives the process is what we see as a cause. The unobservable force acts as the mediating causal process that shapes design.

What is a comparative prediction? A change in some condition, acting as a partial cause, that leads to a change in some effect, mediated by a force, $P \to F \to E$. Comparing different conditions predicts the direction of change in the designed effect.

Why must we use comparative predictions to study design? In practice, inferring cause requires associating the change in one observable with the change in another observable. Inferring cause depends on change, and change arises from comparison.

What is the relation between observable change and inferred cause? Observable change is the story. Inferred cause is a compelling plot to explain the story.

How can comparative predictions mislead? We can observe predicted associations, but our explanations may give the wrong reasons for the observed associations. The best we can do is to test for potential confounding factors and to test over many different conditions.

Are there alternative ways to study the causes of design? Not really. Consistency between an observed story and an explanatory plot is good but leaves open too many alternative consistent plots. Comparison restricts alternatives more strongly than does consistency.[350]

My emphasis on comparison is not new. Darwin often made comparative predictions about how a novel environmental challenge would likely cause an altered design for a matching trait.

Darwin also understood the importance of using phylogenetic history to develop meaningful comparisons.[75] A changed environment might cause all species of a genus to share a novel design. All of the altered species together count as a single change because they share by common descent the same single change in design.

The number of separate comparative observations of change depends on the number of independent events when mapped onto the phylogenetic history.[173]

Darwin's comparative method applies to historical, uncontrolled comparisons. By contrast, the experimental method provides well-controlled comparisons. With proper randomization, we increase the isolation of partial causes. Greater isolation improves the chance that an observed association between a putative cause and a consequent effect is mediated by the hypothesized force rather than by other correlated causes.

Given the highly developed understanding of comparison, why have I emphasized that aspect so strongly? Because most theory about biological design and most studies that discuss the causes of biological design do not focus sufficiently on comparison.

Three reasons may explain the lack of emphasis on comparison. First, biological data are often collected for reasons other than inferring the causes of design. Data often present a pattern rather than how pattern changes with some hypothesized cause. An observed pattern invites a consistent explanation. Consistency greatly dominates over comparison in the literature.

Second, evolutionary theory, which should drive the study of design, often fails to highlight comparative predictions. If a theory does not explicitly conclude and strongly emphasize that, as a condition A changes, the theory predicts that the design feature B changes in a specific way, then that analysis has failed to provide the proper impetus for study.

Third, there is often a mismatch in scale. Forces of design may change relatively quickly. Observed changes in traits may be measured over longer timescales. Microbes provide opportunities to match the scales of force and change. That potential match of scales in microbial studies motivated this book.

There are, of course, many examples of comparative predictions and empirical tests. But progress is slowed by the more numerous studies that emphasize consistency rather than comparison.

Part 1 summarized the forces that, in theory, shape biological design. Those forces lead to comparative hypotheses across many aspects of design and across all forms of life.

In Part 2, I apply the comparative analysis of design to microbial metabolism. To develop that topic, I synthesize the current understanding of microbial metabolism, critique the recent literature's approach to the study of design, and present an improved way to study design by emphasizing comparative predictions and empirical tests.

Part 2

The Design of Metabolism

9 Microbial Metabolism

Catabolic biochemistry extracts free energy from food. Core catabolism arose early in evolutionary history. Modern microbes express many design variants around the basic catabolic core.[208]

Design variants arose for several reasons. Alternative food sources require specialized processing. Disparate environments alter the redox flow of electrons. Different catabolic products provide alternative precursors for building new molecules.

Biochemical innovations change opportunity, leading to novel forces that further alter design. Diversity also arises because different biochemical designs accomplish the same function.

Core catabolism sets a primary challenge in the study of biological design. If we cannot explain design variants in the most basic aspects of microbial biochemistry and energetics, we are not going to succeed for the diverse physiological and behavioral adaptations of life that build on core energetics.

This second part of the book analyzes the design of microbial metabolism. What are the great puzzles in understanding metabolic design? How do we go about solving puzzles of biological design?

To address those questions, I synthesize current knowledge about microbial catabolism and its consequences for microbial fitness. That synthesis highlights puzzles of biochemical design. I then go after my primary goal: solving puzzles of design in biology.

To achieve that goal, I build on the principles in Part 1. My primary method for analyzing design is comparative prediction. Can we predict how a change in some environmental attribute alters metabolic design?

To say that we understand design, we must formulate such comparative predictions and test those predictions. This second part lays the foundation for generating comparative predictions about microbial metabolism. Along the way, I develop many specific predictions.

The methods of approach and the listing of comparative predictions light the way forward in the study of metabolic design. The same approach illuminates design throughout life.

10 Growth Rate

This chapter introduces comparative hypotheses for microbial growth rate. Later chapters develop the underlying thermodynamics, biochemistry, and mechanistic details of metabolism.

The first section restates the problem of analyzing design. Growth rate plays a special role. Designs that grow faster over a sufficiently long period of time dominate life. Determining what is a *sufficiently long period of time* presents a key challenge. Comparative hypotheses about design meet that challenge.

The second section considers how to test comparative hypotheses. The broad range of modern empirical methods provides wide opportunity.

The third section lists comparative hypotheses. The immediate growth rate trades off against other key life history traits, influencing the long-term growth rate and the favored design.

The fourth section focuses on the life history tradeoffs. Empirical study of those tradeoffs can be difficult. I advocate hypotheses that predict how changed conditions alter the relative strength of competing tradeoffs.

10.1 Comparative Hypotheses in the Study of Design

Successful designs increase. *Success* means becoming more common over a sufficiently long period of time, a higher long-term growth rate than the alternatives.

However, we can rarely measure long-term growth rate or identify all of the forces that cause variation in growth rate. To study design, we must seek the most effective ways to gain partial insight. This section briefly reviews Part 1's methods.

EXAMPLE: RATE VERSUS YIELD

To drive metabolic processes faster, additional resources may be required. Resources used to increase growth rate lower the resources available to produce biomass yield. Rate trades off against yield.[317,444]

Suppose a growth cycle begins with a fixed amount of food. Faster growth rate in the short term exhausts the food resources more quickly, reducing the total biomass yield over the full growth cycle.

The long-term growth rate is the total number of descendants (yield) divided by the time period. The rate versus yield tradeoff is an exchange between the short-term and long-term growth rates.

Like most tradeoffs, the opposition of rate and yield expresses a partial truth but also hides other important factors. Figure 10.1 illustrates one potential complication. Progress requires understanding both the power and the limitation of simple tradeoffs in the study of design.

Comparative Predictions

The rate versus yield tradeoff makes a simple comparative prediction. The longer fixed-resource patches last, the more strongly the forces of design favor slower initial growth to achieve higher long-term yield. Equivalently, the shorter resource patches last, the more likely it is that patches disappear before all resources are used up, reducing the cost of rapid inefficient growth and favoring faster initial growth rate (Section 4.1).[130]

We can write the prediction as

$$\text{patch lifespan} \rightarrow \text{marginal yield} \dashv \text{rate}.$$

Increased lifespan of resource patches enhances ($\rightarrow$) the marginal benefit of yield efficiency, which decreases ($\dashv$) the short-term growth rate. Equivalently, when patch lifespan declines and time is short, it pays to grow quickly. Chapter 3 introduced the logic and notation for comparative predictions.

This comparative prediction identifies a simple, powerful force that likely plays an important role in shaping growth rate. However, such predictions depend on *all else being equal*. Of course, all else is not equal. How can we retain the benefit of simple clarity and mitigate the cost of oversimplification?

Focus on General Tendency

If patches disappear before resources are used up, the gain rises for faster short-term growth and lower long-term yield efficiency. Shorter patch lifespans generally favor growing more quickly.

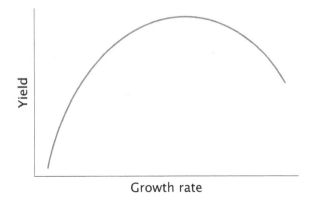

Figure 10.1   Multiple causes influence the rate-yield association. The plot shows an example of a three-way tradeoff between rate, yield, and maintenance. Here, growth rate is $\mu(q)$, with q as food uptake rate. Yield is $y(q) = c\mu(q)/q$, biomass production per unit food uptake, with c as a scaling constant. When food uptake rate q is low, growth increases slowly with q because most re- sources go to maintenance, m. As food uptake increases, more resources are devoted to growth until growth ultimately saturates. We can write an example S-shaped relation between $\mu(q)$ and q as $\mu(q) = (q - m)^a/(4^a + (q - m)^a)$, with $a = 1.7$, $m = 1$, and $q > m$. With these assumptions, growth and yield initially increase together as the relative allocation to maintenance declines. As maintenance becomes a small fraction of the total allocation, the rate-yield tradeoff dominates, causing rate and yield to become negatively associated.[65,322] In lab studies, fast-growing yeast and *E. coli* strains often have a negative associ- ation between growth rate and yield as expressed by the rate-yield tradeoff.[65,326] However, at slower growth rates, a positive association between growth and yield may be observed.[65,288,322]

Confounding causes may obscure that general tendency. For example, short patch lifespan may put a high premium on dispersal. Greater investment in dispersal may reduce growth rate. In that case, shorter patch lifespan may lead to lower observed growth rate.

Observed pattern will not match the general tendency in every case. However, the general tendency should hold when studied over many different conditions. If sufficient randomization of the confounding factors occurs by aggregating over various situations, then the observed tendency for change should match the predicted effect.

Consider Alternative Comparative Predictions

Studying a single comparative prediction in isolation hides confounding causes. Variation in outcome will be high and the signal will be weak.

Listing alternative predictions helps. For patch lifespan and growth rate, we may consider alternatives such as

patch lifespan → marginal yield ⊣ rate

patch lifespan ⊣ marginal dispersal ⊣ rate

patch lifespan ⊣ genetic similarity ⊣ rate.

The first line repeats the prediction that increased patch lifespan raises the marginal benefit of reproductive yield, favoring reduced growth rate.

The second line expresses a tradeoff between dispersal and growth rate. Longer patch lifespan reduces the marginal benefit of dispersal. If dispersal trades off against growth, then less dispersal enhances growth rate. Thus, increasing patch lifespan raises the growth rate.

The third line assumes that longer patch lifespan allows mutations or new immigrants to reduce genetic similarity within patches. Reduced similarity increases competition between genotypes, which favors faster growth rate (Section 5.2). In this pathway of partial causation, increasing patch lifespan raises the growth rate.

With a better sense of potential alternatives, we can more clearly isolate partial causes (Fig. 4.3). In the first pathway, the mediating causal force arises from the marginal valuation of total reproductive yield over the full demographic cycle. We can design comparisons and statistical analyses to separate that explanation from alternative partial causes.

The more complete our description of alternative causal pathways, the more likely we can isolate and study the strength of particular forces that shape design.

This book develops many comparative predictions for microbial traits. Those predictions provide the foundation on which to build alternative causal explanations and empirical tests. New studies will often need additional predictions matched to the particular problem.

Study Comparative Hypotheses about Tradeoffs

Predictions about design often arise from tradeoffs. Longer patch lifespan alters the value of growth rate versus yield efficiency. Shorter patch lifespan alters the value of dispersal versus growth rate.

Tradeoffs may conflict. For example, if resource usage dominates, shorter patch lifespan favors faster growth at the expense of lower yield. If the need to get out of short-lived patches dominates, shorter patch lifespan favors dispersal at the expense of slower growth.

Conflicting tradeoffs and inconsistent predictions make it difficult to test hypotheses. Two approaches help.

First, one can aggregate comparisons over different conditions, as previously noted. Aggregation highlights the average effect. If reduced patch lifespan typically causes growth rate to increase, then resource usage apparently dominates over the need to get out of patches. The rate versus yield tradeoff dominates over the dispersal versus rate tradeoff.

The dominating tradeoff may vary between cases. But the more important effect tends to emerge when aggregated over many conditions.

Second, one can formulate comparative predictions about the tradeoffs. Explicit predictions allow direct testing of tradeoffs.

Consider the three-way maintenance versus growth rate versus yield tradeoff in Fig. 10.1. As resources increase, maintenance allocation becomes less important. The dominating tradeoff shifts from growth rate versus maintenance to growth rate versus yield. In general, one can predict the relative strength of competing tradeoffs as environmental parameters change.

10.2   Testing Comparative Predictions

This section reviews the kinds of empirical data that may be used to test comparative predictions. The potential for comparative tests justifies this book's conceptual framework and specific hypotheses. I illustrate primarily with metabolic traits. I mention a few other traits.

LACTOBACILLI: WHERE THERE IS VARIATION, THERE IS HOPE

Lactobacilli provide agricultural, industrial, and medical value. Their genetics and biology have been analyzed extensively. Some species specialize narrowly on particular carbohydrates and habitats. Other species occur broadly across food sources and ecological niches.[90,264,382]

Lactobacilli's widely varying metabolism and ecology offer good opportunities for comparative study. Prior work sequenced multiple genomes, inferred phylogeny, and associated genetics to metabolism and other key traits.[90]

Martino et al.[264] analyzed 54 *Lactobacillus plantarum* genomes isolated from different environments. Their fine-scale phylogenomic resolution and association of strains with different habitats and metabolic traits laid the foundation for future tests of comparative hypotheses.

SECRETED PROTEINS: COMPARATIVE TESTS IN NATURAL POPULATIONS

Secreted proteins often function as public goods, which are costly to the producer and potentially beneficial to all neighbors. Chapters 3 and 4 introduced public goods and some comparative predictions.

A recent study's abstract summarizes its main conclusions:[149]

> We test the hypothesis that the frequency and cost of extra-cellular proteins produced by bacteria, which often depend on cooperative processes, vary with habitat structure and community diversity. ... [B]acteria living in more structured habitats encode more extracellular proteins. ... Community diversity ... [enhances] proteins implicated in antagonistic interactions and ... [reduces] those involved in nutrient acquisition. Extracellular proteins are costly and endure stronger selective pressure for low cost and for low diffusivity in less structured habitats and in more diverse communities. Finally, bacteria found in multiple types of habitats ... encode more extracellular proteins than niche-restricted bacteria.

This study shows that one can formulate broad comparative hypotheses and test those hypotheses in natural populations. The particular concepts, methods, and conclusions will always be open to debate and improvement. That is the path of progress.

ADDITIONAL COMPARATIVE STUDIES IN NATURAL POPULATIONS

Resource-rich environments associate with high ribosomal number and fast growth.—Increased resources favor higher growth rate and lower biomass yield efficiency,

$$\text{resources} \dashv \text{marginal yield} \dashv \text{rate.}$$

As resources increase, the marginal benefit of yield efficiency declines. Lower gain from yield raises the relative benefit of fast growth. Thus, increasing resources favor faster growth.

Roller et al.[343] associated the number of ribosomal RNA (rRNA) operons with growth rate and yield efficiency in 1167 bacterial species. Maximum growth rate approximately doubled with a doubling of rRNA copy number. Yield efficiency declined with maximal growth rate and rRNA copy number.

The number of ribosomes apparently limits the potential rate of protein production and thus growth rate. The cost of building and running extra ribosomes seems to degrade the biomass yield efficiency per unit of food intake.

Roller et al.[343] argued that resource-poor environments favor genome streamlining and a reduction in costly chemotactic traits. Thus, relatively large genomes and increased chemotaxis indicate resource-rich environments.

Among the 1167 bacterial species, they found significant positive correlations between increased indicators of resources, higher numbers of rRNA operons, faster growth rates, and lower yield efficiencies.

Resource-poor environments emphasize maintenance at the expense of growth and yield.—Laboratory studies often measure physiology under resource-rich conditions. By contrast, many microbes live under resource-poor conditions.[50,185,286,346,367]

To analyze life history under slow growth conditions, Müller et al.[288] studied the methanogen *Methanococcus maripaludis* in low-resource environments. At the most limited resource levels measured, cells devoted 30–50% of catabolic free energy to maintenance. Cells grew slowly and had low biomass yield efficiency.

As food availability rose from that low level, relative allocation to maintenance declined and both growth rate and yield increased. That pattern follows the model in Fig. 10.1.

In contrast with Roller et al.'s[343] observed associations between growth rate, ribosome number, and other cellular attributes, this methanogen maintained relatively constant ribosomal number, cell size, and protein content over wide variations in growth rate.

Müller et al.[288] suggest that different conditions may shift the relative dominance of alternative tradeoffs. For example, high resource environments may favor faster growth rate at the expense of increased maintenance costs for high ribosomal numbers, as indicated by Roller et al.'s[343] study.

Frequent changes in resource level may favor reducing ribosomal maintenance costs in exchange for greater metabolic responsiveness. In particular, maintaining relatively constant ribosome numbers may free resources for a quick growth-rate upshift in response to a rapid increase in resources and a quick growth-rate downshift as resources dissipate.

Finally, low-resource environments may favor trading reduced responsiveness for lower maintenance costs of regulatory control.

It would be interesting to clarify and test comparative predictions for the shifting dominance of alternative tradeoffs as conditions change.

Rare versus dominant oceanic bacterial genomes differ in regulatory complexity and potential for fast growth.—Yooseph et al.[463] annotated metagenomic data from 137 oceanic isolates. Most taxa are typically rare. A few abundant taxa occur widely.

Genomic analysis suggests that rare microbes grow slowly in poor environments and rapidly in rich environments. By contrast, the widely distributed microbes have smaller genomes and seem limited to slow growth rate.

The common, slowly growing taxa have less transcription-regulated control and fewer genes for energy-linked uptake of resources. They also tend to lack genes for chemotaxis, motility, and anaerobic metabolism. Those limitations suggest less capacity for ramping up metabolism and growth in rich environments.

Broad metagenomic surveys suggest potential comparative differences. Subsequent studies could test explicit comparative hypotheses.

Increased resources associate with greater chemical warfare.—In Norway, organic acids increase in snow over the spring months, indicating an increase in microbial populations. Metagenomic analysis of microbial snow communities showed an associated increase in antibiotic resistance genes. The increased antibiotic resistance may be caused by greater chemical warfare associated with rising resource level, microbial density, and competitive interactions.[38]

Habitat nickel concentration associates with a tradeoff between nickel tolerance and growth rate.—Serpentine soil has relatively high nickel abundance.[46] That metal creates a stressful, toxic environment for many microbes.[324] Greater tolerance to toxicity may trade off against other components of success.

Porter & Rice[324] compared metal tolerance and growth rate between serpentine and nonserpentine soils in *Mesorhizobium* bacteria. Greater soil nickel concentration was associated with greater metal tolerance and lower growth rate.

LAB STUDIES OF NATURAL ISOLATES

Lab studies provide opportunity for controlled experiments and focused tests of comparative hypotheses. However, lab studies also change or remove many of the challenges faced by organisms in their natural environments.

The absence of numerous challenges shifts the balance among various tradeoffs that shape design. Thus, lab studies are easier to control and to understand, but they are not necessarily easier to interpret with regard to the forces that shape design. To interpret design, one must combine insights from multiple approaches.

Many lab studies have been published in recent years. This subsection mentions a few studies that analyzed natural isolates. The following subsection describes a few experiments of model lab microbes.

Increased resources and patch lifespan alter functional gene classes.— Song et al.[390] cultured natural soil sample isolates. They grew replicates in different nutrient levels and sampled those replicates after different time periods of growth.

Metagenomic analysis measured the changes in various functional gene classes with changes in nutrient level and growth period. I highlight a few interesting associations.

Low-nutrient environments enriched genes associated with carbohydrate metabolism. Genes associated with antibiotic resistance also increased, perhaps caused by increased competition and greater investment in attack and defense.

High-nutrient environments enriched genes associated with cell division and the cell cycle. Interestingly, dormancy and sporulation genes also increased.

Sampling after a short growth period enriched genes associated with quorum sensing and biofilm formation, toxin-antitoxin systems, and extracellular iron acquisition.

Sampling after a long period of growth enriched genes associated with chemotaxis and cell motility.

These changes may arise from altered species composition rather than evolutionary change within a species. Nonetheless, the selective sorting of functional attributes reveals the forces that shape design.

The authors presented clear comparative hypotheses and experiments to test those hypotheses. This approach provides a useful template for future studies. There will, of course, always be debate about experimental methods, data analysis, and conclusions. My point here concerns the general way in which to form comparative hypotheses and tests.

Increased periods of starvation enhance survival in resource-poor environments.—Baker et al.[22] obtained human saliva isolates. The isolates were first cultured in the lab under high food availability conditions that maintained a broad diversity of species. Then, after adaptation to the culture conditions, the experimenters imposed starvation.

The starvation conditions altered the ecological dominance of species, favoring two *Klebsiella* and one *Providencia* species. The survivors have relatively large genomes, which can catabolize a broader array of food sources than many potential competitors. The survivors also expressed various cyclic depsipeptides that may kill competitors. Those survivors have wide spectra of antibiotic resistance, suggesting broad success in interspecies chemical warfare.

Within the dominant lineages, particular single nucleotide mutations increased in frequency. Some of those mutations occurred in genes previously shown to have a growth advantage in stationary phase, when resources are limited and starvation may be common.

Gene expression patterns also evolved within lineages. For example, competition under starvation increased negative regulation of flagellum motility, negative regulation of biofilm formation, and positive regulation of carbohydrate metabolic processes.

This experiment highlights the significant potential for testing comparative hypotheses. A controlled change in environmental condition leads to strong selective sorting of variant traits. In this particular study, no clear comparative hypothesis was promoted at the start, and the conclusions about particular traits are weak. But the methods suggest great promise.

As annotations of gene sequences to functional classes improve, we will obtain a more refined match between observed nucleotide changes and particular traits. Similarly, better annotations will improve the interpretation of changing gene expression patterns. Those analytical improvements coupled with well formulated comparative hypotheses will provide insight into the forces that shape organismal design.

Random mutagenesis method indicates how changed conditions alter particular genes and traits.—Wu et al.[457] inoculated gnotobiotic mice with 15 human gut bacterial species. They introduced diverse mutations into four of the human-derived bacterial species by transposon-induced mutagenesis. The mice received different diets, defining the environmental change for comparison.

This approach did not test a specific comparative hypothesis. Instead, the broad mutagenesis method provided information about the evolutionary response of diverse genes and associated traits under different dietary conditions.

The introduced mutations created gene-specific markers. Changes in gene frequencies were estimated by comparing the frequencies of the markers after growth with the initial frequencies introduced into the mice.

Dietary differences between the mice altered many nutrients available to the in vivo bacterial community. Thus, one can associate changes in nutrients with differences in evolutionary response measured by gene frequency changes.

Wu et al.[457] also analyzed in vitro experiments of cultured bacteria. Specific nutrients were varied in vitro to evaluate the role of specific factors in altering the frequencies of particular genes.

Finally, this study measured changes in gene expression profiles. The broad measurement of gene frequency and gene expression changes provided detailed information about evolutionary response under different conditions.

This particular experiment focused on dietary changes. Thus, most of the strongly responding genes play a role in the uptake and biochemical processing of food sources.

The same approach could be used to test comparative hypotheses. One could measure in detail how particular demographic and environmental variables alter the evolutionary response of particular traits.

EXPERIMENTAL EVOLUTION OF MODEL SPECIES

Numerous studies evolve microbes under controlled lab conditions.[20,93] Experiments often focus on model species, such as *Escherichia coli* and *Saccharomyces cerevisiae*. I briefly summarize a few examples.

Increased resources favor high growth rate.—Some experiments grow microbes through multiple generations of serial passage. In each generation, a population increases for a period of time and then a sample is transferred to fresh media. If the transfer is done before depleted resources limit growth, then these experimental conditions typically favor increased growth rate.[231,406]

Lewis et al.[239] used metabolic flux models to predict which pathways would be upregulated or downregulated to maximize growth rate. They tested their predictions by comparing gene expression profiles of an *E. coli* strain before and after serial passage that favors increased growth rate. They found a very strong match between their model predictions for changes in gene expression and the observed expression profiles.

Increased growth rate over the fixed time interval in each generation can arise from the increased rate of biochemical reactions or from the greater biomass efficiency yield per unit of resource taken up.[366]

Lewis et al.[239] and prior studies under similar growth conditions[363,406] found evidence supporting both faster reaction rates and increased yield efficiency. Thus, rate versus yield is not the primary tradeoff shaping design under these conditions of excess resources.

Figuring out which tradeoffs dominate under particular conditions is often difficult. Explicit comparative hypotheses about how tradeoffs change in response to changing conditions may be necessary.

Reduced resources favor high yield.—Chemostats provide an alternative to serial passage. Resources flow into the chemostat, and excess cells and waste products flow out. The greater the flow rate, the faster the microbes must grow to balance dilution. The dilution rate expresses the resource flow rate and the approximate steady-state growth rate.

Postma et al.[326] studied *S. cerevisiae* metabolism at various dilution rates. At a low dilution rate below $0.30\,h^{-1}$, associated with low resource flow, input glucose was completely transformed into biomass and CO_2. These conditions lead to a low growth rate and a maximum yield efficiency of $0.50\,g$ of biomass per gram of glucose.

Increased dilution rate up to $0.38\,h^{-1}$ and faster associated growth led to production of acetate, pyruvate, and a reduced biomass yield of $0.47\,g$ per gram of glucose. Further increases in dilution and growth rates triggered aerobic alcoholic fermentation in addition to respiration, with a decrease in biomass yield to a low of $0.16\,g$ per gram of glucose.

Three notable points arise. First, chemostats typically create a steady level of resource limitation. In this particular chemostat experiment, the limiting resource conditions emphasize the rate versus yield tradeoff.

Second, this study primarily measures metabolic changes of cells in response to altered conditions rather than evolutionary changes within lineages. We must consider both the physiological response function of traits to changed conditions and the evolution of the response function as the environment changes. Experiments can test comparative hypotheses about both physiological and evolutionary aspects of change.

Third, interpreting comparisons between widely diverged species may be difficult. The genomic and physiological contexts for the prior bacterial example and this yeast example are so different that comparing rate versus yield aspects between them may provide only limited insight.

Comparisons of parameter changes require considering all else equal. One achieves all else equal for other attributes by matching, controlling, randomizing, or correcting. In the study of biological design, approximating all else equal may require aggregating over various instances of a well-chosen comparison and hoping for partial randomization of other attributes. The meta-trend may be the most useful signal.

More competitors for the same food resource increase growth rate and decrease yield.—Ketola et al.[207] tested this comparative hypothesis by experimental evolution of lab-adapted bacterial species. They grew the focal species *Serratia marcescens* alone or mixed with three other species chosen from a pool of six alternatives.

The focal species evolved a higher growth rate and lower yield when mixed with competitors compared with growing alone. This rate-yield change in response to increased mixtures supports the comparative theory discussed in Section 4.1.

Another study did not observe the same clear increase in growth rate for mixed species versus single species.[226] The experimental setup and the species analyzed differed between the two studies. Ketola et al.[207] propose hypotheses to explain the differing results.

Here, the only important point is that one can test comparative hypotheses about microbial growth and yield by using experimental evolution. Further hypotheses and tests will be needed to clarify why differing results occur in different situations.

More open space for colonization increases motility and decreases growth rate.—Gude et al.[164] measured the changing abundance of two competing *E. coli* strains, labeled A and B. When grown in shaken flasks, strain A greatly increased in frequency relative to B, because A has a higher growth rate in direct competition.

When both A and B are inoculated into a small spot on a soft-agar gel, strain A outcompeted B near the inoculation site. By contrast, strain B dominated beyond the neighborhood of the initial inoculation. These patterns arose because A had a higher growth rate and B had greater motility.

This growth-motility tradeoff led to spatial coexistence when sufficient space existed for B to realize the benefit of its superior dispersal potential. Thus, the more space available for colonization, the more strongly selection favors dispersal rate at the expense of growth rate.

It is common to think of metabolism in terms of biochemistry and the flux through pathways to optimize growth or yield. However, metabolism often trades off with other aspects of success when organisms are considered in their ecological, demographic, and life history context.

The design of metabolism and its biochemical attributes can only be understood within a broad analysis. Comparative tests, such as this growth-motility experimental analysis, will enhance the understanding of design.

Later chapters discuss specific attributes of biochemistry and metabolism that influence growth rate and yield. The ultimate goal is to combine the mechanistic details of particular biochemical pathways with the broader study of comparative life history.

SCALE OF CHANGE IN DESIGN

Plata et al.[323] did a broad comparative analysis of growth across diverse bacterial families. They also analyzed the metabolic phenotypes associated with deletion of particular genes, providing a link between genes and metabolic function.

The broad scope of their study allowed comparison of evolutionary rates at different scales of taxonomic divergence. Among several conclusions (p. 369): "We also find that although a rapid phenotypic evolution is sometimes observed within the same species, a transition from high to low phenotypic similarity occurs primarily at the genus level."

This scaling of evolutionary change calls attention to how one should interpret experimental evolution studies in relation to biological design. In particular, short-term evolutionary response within lineages may perhaps be limited by genetic variation or constrained by mechanism, whereas long-term response associated with higher taxonomic divergence may find ways around those short-term constraints.

In summary, the studies in this section hint at how one might associate changed conditions to changed traits. The examples typically highlight a specific comparison. The studies do not arrive at comprehensively supported conclusions. But they do show the potential for testing comparative hypotheses.

10.3 Comparative Predictions about Growth Rate

This section focuses on growth as an abstract trait, emphasizing general concepts. I include predictions previously mentioned. The next section introduces comparative predictions about tradeoffs between growth rate and other life history traits.

These abstract predictions about growth rate complete the introductory overview of metabolism in relation to life history. The following chapters develop predictions about growth based on mechanistic aspects of biochemistry and metabolism.

GENERAL FACTORS

These predictions give a sense of how an isolated force may influence design. In any real application, one must adapt the predictions to account for the variety of forces acting simultaneously and the opportunities for isolating individual forces.

Mixing ⊣ similarity ⊣ rate. Increased genetic mixing lowers the genetic similarity (relatedness) between neighbors. Lower similarity favors faster growth rate to outcompete neighbors, enhancing relative genetic representation in future generations. When similarity is high, outgrowing neighbors provides little gain, reducing the benefit of fast growth.[130,317]

Resources ⊣ marginal yield benefit ⊣ rate. Increased resources reduce the marginal gain from efficient yield conversion of resources into reproductive output. Lower gain from yield efficiency raises the relative benefit of fast growth under a rate-yield tradeoff. Thus, increasing resources favor faster growth.

Resources ⊣ marginal dispersal benefit ⊣ rate. Increased resources reduce the marginal benefit of dispersal. Lower investment in dispersal raises the growth rate.

Resources ⊣ marginal attack benefit ⊣ rate. Increased resources lower the marginal benefit of killing competitors. Less investment in attack raises the growth rate.

Attack → marginal defense benefit ⊣ rate. Increased attack by neighbors raises the marginal benefit of defense. More investment devoted to defense lowers the growth rate.

Timescale and Demography

Immediate growth rate may trade off against long-term success. For example, growing faster in the short term may lead to reduced overall yield and production of descendants in the long term. Long-term success is a measure of growth rate on a longer timescale: the total number of descendants divided by a relatively long total period of time.

What is the proper timescale over which to understand organismal design? That question is among the most important and most difficult problems in analyzing design.

If some particular design has the highest immediate growth rate, then that design will be increasing in frequency and becoming more dominant in the short term. If that same design has relatively low growth rate over long time periods, then that design will be decreasing and less prevalent over that longer time scale.

A timescale between the instantaneous and the infinite must dominate the designs that we observe. For the problems that I focus on in this book, the demographic life cycle often provides a convenient time unit.[122]

Suppose lineages colonize resource patches, grow on those patches, disperse descendants from those patches to colonize new patches, and then die within their natal patches.

Designs with the highest growth rate over a full cycle will tend to increase and dominate. A high long-term success over the full cycle may associate with relatively low instantaneous growth rate by, for example, a tradeoff between the immediate growth rate and the long-term yield.

Various tradeoffs may create a similar opposition between immediate growth rate and long-term success. Comparative predictions follow.

Patch lifespan → marginal yield ⊣ rate. The longer a patch lasts, the more likely resources will be depleted and the more strongly total reproduction depends on yield efficiency. A rate-yield tradeoff may associate higher long-term yield with lower immediate growth rate.

Patch lifespan ⊣ similarity ⊣ rate. The longer a patch lasts, the more likely mutation or additional genetic mixing reduces similarity among neighbors. Reduced similarity favors faster growth to outcompete unrelated neighbors.

Patch lifespan ⊣ marginal dispersal benefit ⊣ rate. The shorter a patch lasts, the more quickly lineages must disperse to colonize other patches. More investment in dispersal may reduce immediate growth rate.

Heterogeneity → evolvability ⊣ rate. Heterogeneous and unpredictable environments often favor a faster rate of adaptive evolution. Evolvability traits include greater recombination, higher mutation rate, and more stochasticity in trait expression. Greater evolvability provides a long-term benefit that may trade off against immediate growth rate.[308,311]

10.4 Comparative Predictions about Tradeoffs

Tradeoffs vary with context, making them difficult to study. One gains the most insight by comparatively predicting how changes in context alter the relative strength of different tradeoffs. In particular, changing conditions strengthen some tradeoffs and weaken others. Testing comparative predictions reveals the shifting dominance of alternative tradeoffs.

Resources ⊣ marginal changes → rate vs yield. Greater resources saturate growth, reducing further marginal changes in growth and yield. Reduced marginal changes weaken the growth-yield tradeoff.

Patch lifespan → marginal changes → rate vs yield. Longer patch lifespan increases the chance of resource limitation. More limited resources enhance marginal cost and benefit changes, increasing the strength of the rate versus yield tradeoff.

Resources ⊣ marginal gains → rate vs dispersal. High resource levels saturate growth and limit the benefits of dispersal, weakening the tradeoff between growth rate and dispersal.

Patch lifespan ⊣ marginal gains → rate vs dispersal. Shorter patch lifespan increases the marginal gains of growth and dispersal, strengthening the tradeoff between growth and dispersal.

Resources ⊣ warfare benefit → rate vs attack–defense. An increase in resources reduces the benefit of killing neighbors to gain a competitive advantage. Smaller gains in warfare decrease the intensity of tradeoffs between growth rate and attack or defense traits.

Patch lifespan → warfare benefit → rate vs attack–defense. Increased patch lifespan may associate with a greater chance of resource limitation. More limited resources enhance the benefits of warfare and increase the intensity of tradeoffs between growth rate and attack or defense traits.

Heterogeneity → evolvability gain → rate vs evolvability. Unpredictable environments enhance the gain from evolvability. Those greater gains may strengthen the tradeoff between long-term evolvability and immediate growth rate.

This chapter's predictions provide very rough qualitative guides. In any application, this initial list helps to locate a starting point, to see what is missing, and to advance by refined assumptions and further analysis. Past empirical studies illustrate the potential methods that may be used to test the honed comparative predictions.

11 Thermodynamics: Biochemical Flux

> The general struggle for existence of animate beings is therefore not a struggle for raw materials ... nor for energy ... but a struggle for entropy.
>
> —Ludwig Boltzmann[43]

> What organisms feed on is negative entropy.
>
> —Erwin Schrödinger[362]

Prior chapters emphasized abstract problems. How do changing environmental factors alter the fitness costs and benefits of traits? How do those fitness costs and benefits shape organismal design?

The remaining chapters link those abstract questions to the biochemistry of microbial metabolism. New comparative predictions follow.

This chapter reviews thermodynamics. The first section begins with free energy, the force that drives biochemical reactions. Free energy measures the production of entropy. Negative entropy in food provides the source for the catabolic increase in entropy, fueling life.

The second section analyzes metabolic reaction rates. Enzymes enhance biochemical flux by lowering the free energy barrier of intermediates. The lowered barrier reduces the resistance against reaction. Biochemical flux equals the free energy driving force divided by the resistance barrier against flux.

The third section links biochemical flux to design. Increased driving force trades off greater flux against lower metabolic efficiency. Reduced resistance trades off greater flux against the cost of catalysis. The costs and benefits of adjusting force or resistance shape metabolism.

Subsequent chapters build on these principles to develop comparative predictions for metabolic design.

11.1 Entropy Production

> We have constantly stressed that thermodynamic systems do
> *not* tend toward states of lower energy. Therefore the ten-
> dency to fall to lower free energy must not be interpreted
> literally in terms of falling down in energy. *The Universe falls
> upward in entropy:* that is the only law of spontaneous change.
> The free energy is, in fact, just a disguised form of the total
> entropy of the Universe ... even though it carries the name
> "energy."
>
> —Peter Atkins[18]

Textbooks emphasize that free energy drives chemical reactions. A reaction proceeds if the inputs have greater free energy than the outputs. The greater the drop in free energy, the faster the reaction will be. Organisms must acquire free energy from their food and use up that free energy to drive their metabolic processes.

The confusion arises because free energy does not measure energy, it measures entropy. Total energy is always conserved. The cause of change is always an increase in total entropy.

In many chemical applications, it is sufficient to calculate free energy without understanding its meaning. But a clear conceptual view helps to understand broader problems of metabolic design.

This section reviews entropy production, the force that drives chemical reactions. Sections 11.2 and 11.3 link entropy change to tradeoffs between reaction rate and metabolic efficiency.

Change in Free Energy Measures Change in Entropy

> [T]he natural cooling of a hot body to the temperature of
> its environment can be readily accounted for by the jostling,
> purposeless wandering of atoms and quanta that we call the
> dispersal of energy. ...All chemical reactions are elaborations
> of cooling, even those that power the body and the brain.
>
> —Peter Atkins[18]

Food provides chemical bonds with concentrated energy. Breaking the food's chemical bonds disperses the energy. Dispersing energy increases entropy. Increasing entropy is a general kind of cooling.

We can think of the chemical bonds in food as storing negative entropy. The amount of negative entropy is the potential to produce entropy by dispersing the energy contained in the chemical bonds. Organisms feed on negative entropy.[362]

To analyze metabolism, we must track the flow of entropy between the chemical reactions and the environment. In chemistry, we typically partition the world into the system on which we focus and the surrounding world. We call the surrounding world the *bath* because we think of the system as an isolated container floating in a large bath with constant temperature and pressure.

If the system produces heat, the bath soaks up and disperses that heat. The temperature in the system remains constant at the external bath's temperature. If the system cools, then the bath transmits heat to the system, maintaining uniform temperature. Similarly, pressure in the system equilibrates with the large external bath.

The total entropy is the bath's entropy plus the system's entropy,

$$S_t = S_b + S_s.$$

By the Second Law of thermodynamics, the change in entropy is never negative,

$$\Delta S_t = \Delta S_b + \Delta S_s \geq 0. \tag{11.1}$$

In this setup, the system exchanges heat energy with the bath. Total energy never changes. But energy can change location or form.

The dispersal of heat energy from the system to the bath, ΔH_b, moves concentrated energy from the system to dispersed energy in the bath.

Energy is more dispersed in the bath because the bath is very large and disperses the energy widely, with essentially no change in temperature. As heat energy disperses, it cools and increases in entropy.

The transfer of heat energy to the bath and its dispersal raises the entropy of the bath by

$$\Delta S_b = \frac{\Delta H_b}{T},$$

in which T is temperature.

Combining the prior equations yields the total change in entropy,

$$\Delta S_t = \frac{\Delta H_b}{T} + \Delta S_s \geq 0. \tag{11.2}$$

A reaction proceeds when total entropy increases, $\Delta S_t > 0$. The change in total entropy depends on the heat produced by the reaction and dissipated through the bath, $\Delta H_b/T$, plus any change in the internal entropy of the system, ΔS_s.

The heat term is negative when a reaction sucks up heat, causing heat to flow from the bath to the system. The internal system entropy term is negative when the reaction creates products with less entropy than the initial reactants. The reaction proceeds only when the overall total entropy increases.

The entropy expression in eqn 11.2 is equivalent to the classic expression for Gibbs free energy used in all textbooks on chemistry. I show the equivalence because it is important to understand that entropy is the ultimate driver for the reactions of metabolism.

I noted above that the transfer of heat energy to the bath raises the entropy of the bath by $\Delta S_b = \Delta H_b/T$. Energy has units in joules, J. Entropy has units JK^{-1}, in which K is temperature in kelvins. Thus, we can write the change in heat energy in terms of the change in entropy as $\Delta H = T\Delta S$.

Because energy is conserved, the change in the heat energy of the bath must be equal and opposite to the change in the heat energy of the system. Thus $\Delta H_b = -\Delta H_s$, and

$$\Delta S_b = -\frac{\Delta H_s}{T}. \tag{11.3}$$

We can combine eqns 11.1 and 11.3 to write the total change in entropy required for a reaction to proceed as

$$\Delta S_t = -\frac{\Delta H_s}{T} + \Delta S_s > 0. \tag{11.4}$$

Multiplying by $-T$ changes the units of the terms to energy and reverses the direction of change required for a reaction to proceed,

$$\Delta G = \Delta H_s - T \Delta S_s < 0, \tag{11.5}$$

in which $\Delta G = -T \Delta S_t$ is the change in the Gibbs free energy. The word *energy* is misleading because total energy can never change. The change in Gibbs free energy quantifies the energy that is free and available to do work, not the change in energy.

A decrease in free energy is actually an increase in entropy. It is the increase in entropy that determines the capacity to do work.

For example, concentrated energy in a chemical bond can be relatively "hot" compared with its surrounding environment. As the "heat" disperses and cools, the heat flow can potentially be captured to do work. Dispersed energy is cold and cannot do work.

An ordered system tends to increase in entropy toward its naturally disordered maximum entropy equilibrium. Work gets done when one can capture the dissipation of an ordered disequilibrium in one system to drive the increasingly ordered disequilibrium of another system.

THE ESSENTIAL COUPLING OF DISEQUILIBRIA

> The central lesson of the Second Law is that natural processes are accompanied by an increase in the entropy of the Universe. A coupling of two processes may cause one of them to go in an unnatural direction if enough chaos is generated by the other to increase the chaos of the world overall.
>
> —Peter Atkins[18]

Metabolism couples chemical reactions. One reaction moves a system toward its equilibrium, producing entropy as it dissipates the disequilibrium. The partner reaction is driven away from its natural equilibrium, decreasing entropy.

As digestion breaks down the chemical bonds in food, the increase in entropy can be coupled to reactions that drive the production of ATP. The increasing disequilibrium of ATP against ADP stores negative entropy in a usable form, a battery that can be tapped to do work.

Metabolic reactions dissipate the ATP–ADP disequilibrium to increase entropy. That increase in entropy drives the coupled entropy-decreasing reactions that build the ordered molecules needed for growth and other life processes.[47,86,181,182,201]

Entropy-producing and entropy-decreasing reactions must be directly and necessarily coupled. The paired reactions proceed only if their combined change produces entropy. Understanding the molecular escapement mechanisms that couple reactions poses one of the great problems of modern biology.[47,58,59]

Consider two coupled reactions. The first reaction breaks down a food molecule into two parts,

$$\text{Food} \longrightarrow \text{PartA} + \text{PartB},$$

producing entropy $\Delta S_1 > 0$. The second reaction drives the production of ATP, increasing the disequilibrium of ATP relative to its components ADP and P_i, as

$$\text{ADP} + P_i \longrightarrow \text{ATP}.$$

Increasing disequilibrium reduces entropy, $\Delta S_2 < 0$, or, equivalently, increases negative entropy. The paired reactions when coupled should be thought of as a single process that, mechanistically, necessarily combines the two half-reactions,

$$\text{Food} + \text{ADP} + P_i \longrightarrow \text{PartA} + \text{PartB} + \text{ATP}.$$

The process proceeds if the total entropy change is greater than zero, $\Delta S_1 + \Delta S_2 > 0$, or equivalently, $\Delta S_1 > -\Delta S_2$. Here, $-\Delta S_2$, the negative entropy captured in the driven reaction producing ATP, must be less than ΔS_1, the entropy produced in the driver reaction digesting the food.

If the reactions are not coupled, the second reaction producing ATP cannot proceed. Instead, the first reaction disperses the concentrated energy in the food, increasing entropy by ΔS_1. Typically, that entropy increase associates with the dissipation of heat. The heat spreads out through the bath and is lost.

When the reactions are coupled, some of the entropy produced by the driver reaction is stored as negative entropy in the driven reaction. The fraction of entropy captured by the driven reaction provides a measure of efficiency,

$$\frac{-\Delta S_2}{\Delta S_1} < 1. \tag{11.6}$$

Because the total entropy produced by the combined reactions must be greater than zero, $\Delta S_1 + \Delta S_2 > 0$, some of the entropy produced must escape unused. In every metabolic reaction, the organism loses some of its negative entropy.

ENTROPY CHANGE IN THE DISSIPATION OF DISEQUILIBRIUM

Entropy change drives reactions. Total entropy change arises from the combined move toward equilibrium in the driver reaction and move away from equilibrium in the driven reaction. This subsection describes the relation between entropy and equilibrium. I use ATP as an example.

The terminal phosphate bond in ATP is commonly described as a high-energy bond. However, when ATP is at equilibrium with its components, the forward and back reactions

$$ADP + P_i \longleftrightarrow ATP$$

happen at equal rates. There is no change in entropy or free energy in either direction.

ATP at equilibrium with ADP cannot drive other reactions. It is not the energy in the phosphate bond that drives reactions. Instead, it is the dissipation of the ATP–ADP disequilibrium toward its equilibrium that drives coupled reactions away from their equilibrium.

The classic textbook expression for free energy in terms of disequilibrium highlights the key concepts. Consider the reaction

$$A + B \longleftrightarrow C.$$

As a reaction proceeds from its initial concentrations toward its equilibrium, the change in free energy is proportional to

$$\Delta G \propto \log \frac{Q}{K}. \tag{11.7}$$

The reaction quotient, Q, is the ratio of the product concentration to the reactant concentrations,

$$Q = \frac{[C]}{[A][B]}.$$

The equilibrium constant, K, expresses the reaction quotient at equilibrium,

$$K = \frac{[C]_{eq}}{[A]_{eq}[B]_{eq}}.$$

The expression $\log Q/K$ measures disequilibrium. Thus, the change in free energy measures the size of the initial disequilibrium that dissipates as the system moves to its equilibrium.

The greater the initial disequilibrium of a driver reaction, the more free energy (negative entropy) is available to drive other reactions.

The further a driven reaction is pushed away its equilibrium, the more free energy must be expended by the driving reaction.

At equilibrium, a system has no free energy to drive other reactions, no matter how much energy is concentrated in its chemical bonds.

THERMODYNAMIC DRIVING FORCE AND REACTION RATE

This subsection relates the change in free energy to the rate of reaction. The key point is that faster reaction rate trades off against lower efficiency. Reduced efficiency means more free energy is dissipated as heat, decreasing the free energy available to drive other reactions.

A negative change in free energy drives a reaction forward, toward its equilibrium. The greater the decrease in free energy, the faster the forward reaction proceeds relative to the reverse reaction,

$$-\Delta G \propto \log \frac{J^+}{J^-}, \tag{11.8}$$

in which J^+ is the forward reaction flux, and J^- is the reverse reaction flux. The decrease in free energy, $-\Delta G$, is the driving force of the reaction.

Section 11.2 links the driving force to the overall reaction rate. Here, I focus on the consequences of a greater driving force for the associated flux ratio, J^+/J^-. An increased flux ratio corresponds to a faster net forward reaction rate but does not tell us the overall rate.

Greater loss of free energy, $-\Delta G$, increases the flux ratio, J^+/J^-, and drives the net forward reaction faster. Loss of free energy relates to gain in entropy. The additional entropy increase corresponds to a greater dissipation of concentrated and ordered energy. Dissipated energy typically flows away as lost heat, which cannot be used to do work.

In other words, an increased reaction rate burns more fuel. Here, *burning fuel* means dissipating negative entropy as heat. That lost heat cannot drive other reactions or do work.

Typically, organisms couple driving reactions that increase entropy by ΔS_1 to driven processes that decrease entropy by $-\Delta S_2$. In eqn 11.6, the efficiency $-\Delta S_2/\Delta S_1$ expresses the fraction of the entropy produced by the driver reaction that is captured by the driven reaction in useful work or in the building of ordered molecules.

These results link rate and efficiency. For coupled reactions, the closer to zero the total amount of entropy produced, $\Delta S_t = \Delta S_1 + \Delta S_2 > 0$, the more efficiently the driven process captures the available potential from the driver reaction.

Greater efficiency means that $-\Delta S_2$ is closer to ΔS_1. The smaller the difference, the less the total free energy decrease of the coupled reactions,

$$-\Delta G_t = T\Delta S_t = T(\Delta S_1 + \Delta S_2) \propto \log \frac{J^+}{J^-}.$$

The smaller the free energy decrease, the slower the net forward reaction flux tends to be.

Maximum efficiency occurs as the entropy captured by the driven reaction approaches the entropy produced by the driver reaction, $-\Delta S_2 \rightarrow \Delta S_1$, causing $\Delta G_t \rightarrow 0$. As the free energy change becomes small and efficiency increases, the net forward reaction flux of the overall coupled reaction declines toward zero. Reaction rate trades off against efficiency.

FREE ENERGY AND ENTROPY

As noted below eqn 11.5, the total entropy change and the free energy change express the same quantity in slightly different ways as

$$\Delta S_t = -\frac{\Delta G}{T}.$$

The standard in chemistry is to use the term *free energy* when discussing entropy changes in reactions.

Free energy is a useful perspective because the quantity is the degree to which energy is concentrated or ordered and, through the dissipation or disordering of that energy, work can be done. The work may be used to drive other reactions or to do physical work.

Because *free energy* is the standard term, and also a useful one, I will switch freely between that term and the more causally meaningful entropy descriptions.

11.2   Force and Resistance Determine Flux

Analyses of biochemical reactions often focus on free energy change as the thermodynamic driving force. However, biochemical flux also depends on the resistance that acts against reactions,

$$\text{flux} = \frac{\text{force}}{\text{resistance}}. \tag{11.9}$$

Biochemical flux is analogous to electric flux in Ohm's law, in which the electric flux (current) equals the electric potential force (voltage) divided by the resistance.[215,314] High voltage with no wire connecting the poles produces no current. A chemical reaction with a large driving force but high resistance proceeds slowly or not at all.

For example, combining hydrogen and oxygen yields water. The reaction causes a large drop in free energy and thus has a powerful thermodynamic driving force. Yet, at standard ambient temperature, nothing happens when one mixes the two gases.

The oxygen-hydrogen reaction intermediate significantly reduces entropy at ambient temperature and thus does not easily form. A reaction intermediate with a large reduction in entropy corresponds to a high activation energy, creating a resistant barrier to reaction.

Organisms modulate chemical flux by altering resistance or force. Catalysts reduce resistance, which increases flux by lowering the activation energy. Adding reactants or removing products raises the forward driving force, which enhances forward flux.

Changing the flux of a reaction alters the tendency for molecular transformation versus stability. Metabolic transformation moves the negative entropy in food into the highly ordered molecules needed for life. Stability protects those useful ordered forms from their intrinsic tendency to decay. Organisms control transformation and stability by modulating the driving force and the resistance of reactions.

11.3   Mechanisms of Metabolic Flux Control

Three mechanisms alter metabolic flux.[100] Each mechanism creates tradeoffs between flux and efficiency. Later chapters show how those tradeoffs shape metabolic design (Fig. 11.1).

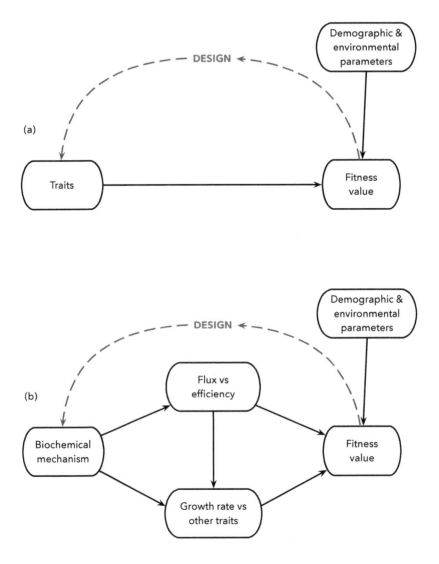

Figure 11.1   Relating mechanism to the study of organismal design. (a) The abstract problem and associated comparative predictions. Traits influence reproduction and fitness. Demographic and environmental parameters alter the fitness value of traits. Fitness value feeds back by natural selection to change traits over time, influencing organismal design. (b) For biochemical aspects of metabolism, we may consider particular mechanisms that influence trait expression and the associated tradeoffs that follow.

SHORT TIMESCALE

On short timescales, changing the metabolite concentrations alters the free energy driving force. When the initial driving force is weak, modest changes in metabolite concentrations can significantly increase the driving force. Greater driving force increases flux and reduces efficiency. Lower efficiency means greater dissipation of metabolic heat and less entropic driving force available to do beneficial work.

Lost benefits include storing less negative entropy in an ATP–ADP disequilibrium, building fewer useful molecules, or doing less physical work by molecular motors.

By contrast, altering metabolite concentrations to lower the driving force enhances efficiency, potentially providing more free energy to build molecules or do physical work. However, the extra usable free energy typically comes more slowly because low driving force associates with reduced reaction rate.

INTERMEDIATE TIMESCALE

On intermediate timescales, modifying enzymes alters the resistance against reactions. Organisms modify enzyme molecules by adding, removing, or changing small pieces. Such covalent modifications of existing enzymes can change their catalytic action. The catalytic change in reaction rate may increase or decrease resistance, providing a mechanism to control metabolic flux via reaction kinetics.

Cells often modulate reactions by covalent changes, catalyzed by widely deployed enzymes. Covalent changes may be faster and the costs of modification may be lower than controlling reaction resistance by building or destroying enzymes.

LONG TIMESCALE

On long timescales, synthesis and degradation change enzyme concentrations. More enzyme lowers resistance and increases flux. Enzymes may be complex molecules that are costly to build and slow to deploy.

For reactions with low driving force, significantly increasing net flux requires a large change in resistance and thus a large and costly change in enzyme concentration. When possible, changes in metabolite concentrations may be more effective and less costly.

For reactions with high driving force, small changes in enzyme concentration significantly alter net flux. That sensitivity to enzyme concentration allows rapid control of flux. However, the high driving force is metabolically inefficient, typically providing benefit only when conditions favor rapid flux.

Metabolite concentration and enzyme activity control force, resistance, and flux. Other mechanisms may affect flux. For example, the surrounding environment in which metabolites float influences entropy, diffusion, interference, and reactivity.[86]

Alternative control mechanisms have different consequences for flux and timescale of action. Costs and benefits arise from changes in flux, speed of adjustment, and thermodynamic efficiency.

12 Flux Modulation: Driving Force

Organisms control metabolic flux by altering the thermodynamic driving force of biochemical reactions. Organisms also modulate flux by altering the resistance that impedes reactions. This chapter focuses on thermodynamic driving force. The next chapter considers resistance.

The total driving force depends on the initial food inputs and the final products. Foods rich in free energy increase the potential driving force. Final products made by passing electrons to strong attractors have low free energy, increasing the potential driving force.

The metabolic cascade flows through many reaction steps. The driving forces for the individual steps sum to the total driving force. Dividing the total among the individual steps sets a key challenge in metabolic design.

For example, low driving force and a slow reaction in one step impede flux through the metabolic cascade. By contrast, high driving force and a fast reaction dissipate a lot of free energy.

Free energy dissipated in one reaction must be balanced by reduced driving force or lower metabolic efficiency for other reactions. The metabolic cascade may fail if there is not enough remaining free energy to drive all reactions.

The first section analyzes glycolysis. Recent advances measure the in vivo driving force for individual reaction steps. Those data illuminate how metabolic design modulates flux through the glycolytic cascade.

The second section discusses overflow metabolism. When a metabolic cascade runs too fast, product concentrations build up, slowing or reversing key reaction steps. To relieve product inhibition, organisms may excrete the inhibiting reaction products.

The excreted products contain usable free energy. Thus, fast metabolic rate trades off against reduced efficiency. That tradeoff provides an excellent model to study the forces that shape metabolic design.

The third section analyzes puzzles of design posed by overflow metabolism. Biochemical mechanisms such as product inhibition explain why excreting usable free energy may happen. But those mechanistic aspects do not explain why cells sometimes grow fast and wastefully excrete resources, while at other times cells grow more slowly and efficiently. Environmental factors that alter fitness costs and benefits ultimately determine design within the constraints imposed by biophysics.

The fourth section describes the alternative timescales for the evolutionary analysis of design. Short-term lab studies typically focus on how changed conditions alter physiological responses. Those studies often reveal biophysical constraints, such as how limited numbers of proteins in cells or limited membrane space constrain design.

Studies over several generations reveal how genetic variation provides opportunity to alter design. In the medium term, altered design typically occurs within the context of the current physiological system.

Comparing species or higher taxonomic levels reveals long-term evolutionary changes in design, including those that modulate biophysical constraints or alter the core physiological system. Such comparisons associate varying environmental challenges with the varying design of organismal traits. Overall, the different studies of metabolic design must be understood in terms of their evolutionary timescales of analysis.

The final section considers alternative glycolytic pathways. The pathways vary in their driving force, in the ways that they capture and store free energy, in the costs of running the cascades, and in the biochemical benefits that they provide for other functions. Those variations raise interesting puzzles of metabolic design.

12.1   Near-Equilibrium Glycolysis

Reactions with small free energy change are near equilibrium and proceed slowly (eqn 11.8). Small free energy change means that, after a reaction occurs, the system retains most of its initial free energy. In a reaction cascade, the retained negative entropy in one step can often be used to drive other reaction steps.

Reactions near equilibrium also have the benefit of easy flux modulation. Small changes in metabolite concentrations push the reaction away from equilibrium. Deviation from equilibrium increases the force that drives the system back toward equilibrium, increasing the reaction rate.

Thus, reactions near equilibrium save negative entropy and easily modulate flux. Those advantages suggest that certain environments favor metabolic reactions to be regulated near equilibrium. However, it has been difficult to measure the in vivo flux and free energy change.

Recent technical advances label metabolites to measure both forward and backward fluxes.[309,310,459] From eqn 11.8, repeated here,

$$-\Delta G \propto \log \frac{J^+}{J^-},$$

measurement of flux in both directions specifies the free energy change and deviation from equilibrium.

Measurements in a single environment can be difficult to interpret. For example, high flux through one reaction and slow near-equilibrium flux through another reaction may reflect the particular conditions rather than a general attribute of system design. Comparison between environments provides more insight.

E. COLI UNDER CHANGING NITROGEN AVAILABILITY

Park et al.[310] compared E. coli flux when grown in limited and abundant nitrogen conditions. Cells grew slowly with arginine as the sole nitrogen source. After adding ammonia, a better nitrogen source, growth increased within minutes by 170% and glucose uptake increased by 60%. The concentrations of glycolytic intermediates did not change much, posing the puzzle of how glycolysis keeps up with increased overall flux.

Under limited nitrogen, many steps of glycolysis were near equilibrium, with small driving force per reaction. Park et al. measured five glycolytic transitions between the uptake of glucose and the output of phosphoenolpyruvate (PEP) near the end of glycolysis, just before pyruvate production (Fig. 12.1).

The overall driving force was low between glucose input and PEP output near the glycolytic endpoint. Most of the free energy change occurred at the first input and final output reactions. The low overall driving force when nitrogen is limited reflects low thermodynamic push from slow glucose uptake and low pull on PEP from limited cellular demand for growth.

Five minutes after adding ammonia to provide more nitrogen, the overall driving force approximately doubled. Most of the increase arose by greater push from rising glucose uptake and greater pull from increased growth demand for PEP and downstream products.

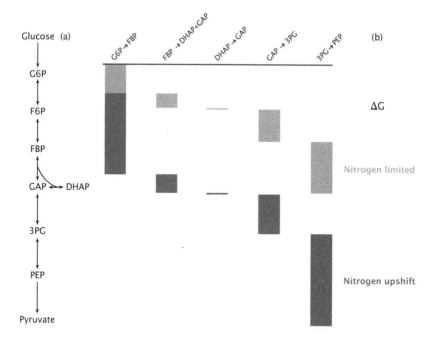

Figure 12.1 Free energy change (ΔG) between various intermediates in the *E. coli* glycolytic pathway. (a) Some of the intermediate molecules between glucose input and pyruvate output. The double-headed arrows indicate empirical measurements for forward and reverse flux for those reversible transformations. Most introductory biology books and biochemistry texts describe details for the full glycolytic cascade. (b) The height of each rectangle shows the relative free energy change between intermediates. The cumulative change is the sum of the changes for intermediate components. Light gray indicates growth under limited nitrogen conditions and dark gray indicates growth under abundant nitrogen. For the first difference on the left, the change for nitrogen upshift combines the heights of both rectangles. Redrawn from Fig. 3 of Park et al.[310]

The intermediate steps also increased their driving force and flux to keep up with demand. Because the reactions were initially close to equilibrium, significant increases in driving force and net flux arose from relatively small changes in concentrations.

The initial near-equilibrium state and rapid response to reactant concentrations require high enzyme concentrations. If enzymes were at low concentrations and nearly saturated, then greater incoming flux of reactants would only partially increase reaction rates.

The excess of enzymes near equilibrium adds a cost. Park et al.[310] conclude that the benefit of being able to increase flux rapidly on nitrogen upshift by small changes in metabolite concentrations provides an overriding benefit. With excess enzyme, increased flux demand can be met by raising the net forward flux per enzyme molecule.

If, by contrast, the system reduced enzyme concentrations at low growth conditions, then, upon nitrogen upshift, several different enzymes would have to be produced simultaneously to raise flux.

Park et al.[310] also studied response to phosphorous upshift. Glucose uptake increased by approximately four-fold, overall driving force increased by about a quarter, and growth rate increased.

When compared to nitrogen upshift, the change in phosphorus is associated with different initial and final driving forces for individual reactions. However, the overall flux increase once again arose primarily through greater driving force, although some enzymatic changes also occurred that lowered the resistance of reactions and increased kinetics.

E. COLI RESPONSE TO OXYGEN UPSHIFT

An upshift in oxygen availability causes a different pattern of change when compared to increases in nitrogen and phosphorus. When oxygen is limited, catabolic processing may end with glycolysis. As oxygen concentration increases, greater flux through the subsequent TCA cycle and electron transport becomes possible (Fig. 12.2).

Oxygen provides a strong electron acceptor that creates much greater overall thermodynamic driving force from the initial uptake of glucose to the final production of water and carbon dioxide. The enhanced driving force enables much larger ATP generation per glucose molecule than from glycolysis alone.

When Park et al.[310] increased oxygen availability, cells decreased glucose consumption and increased growth rate, consistent with greater extraction of usable negative entropy per unit of food input. Additional oxygen slowed glycolysis, which likely resulted in spare enzymatic capacity in the glycolytic steps, associated with a move toward equilibrium and closer balance of forward and backward fluxes.

From a highly oxygenated state, a new increase in limiting nitrogen or phosphorus could once again use the spare enzyme capacity of near-equilibrium reactions to trigger a very rapid glycolytic increase.

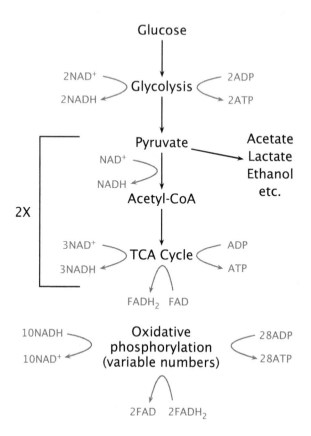

Figure 12.2   Rough sketch of the major catabolic pathways. Organisms vary in many details, including the numbers of ATP, NADH, and FADH$_2$ produced and consumed. Foods composed of proteins or lipids are catabolized through different initial pathways, typically producing pyruvate or acetyl-CoA. Cells may excrete acetate, lactate, ethanol, or other glycolytic products rather than pass those products through acetyl-CoA and the TCA cycle. Glycolysis produces two pyruvate molecules, doubling the stochiometry of the lower pathways relative to the initial glucose input. Electron transport and oxidative phosphorylation use variable numbers of input NADH and FADH$_2$ and convert a variable number of ADP to ATP. For example, one glucose and the consequent two pyruvates may associate with approximately 10 NADH and 2 FADH$_2$ inputs and 28 ATP outputs. Adding the 2 ATP from glycolysis and the 2 ATP from the 2 TCA cycles, a single glucose molecule may yield approximately 32 ATP.[428]

Differences between Species in Anaerobic Glycolytic Driving Force and Free Energy Efficiency

When grown on glucose and with abundant nutrients, the anaerobic cellulose digesters *Clostridium acetobutylicum* and *Thermoanaerobacterium saccharolyticum* have strong glycolytic driving force and rapid growth similar to anaerobically grown *E. coli*.

By comparison, the cellulose digesters *C. cellulolyticum* and *C. thermocellum* grown on glucose have glycolytic driving force reduced by 80–90% and slower growth. Because these species dissipate less free energy across glycolysis, they can potentially capture more usable negative entropy per glucose molecule, obtained as more ATP generated per glucose input.[193,310]

Figure 12.3 illustrates the differences between the fast and slow species.

Puzzles of Design

Why do species vary so much in glycolytic driving force and associated efficiency in extracting and storing free energy? Why do species such as *E. coli* respond to increased nutrients by rapidly enhancing glycolytic free energy change, flux, and growth, whereas species such as *C. cellulolyticum* retain low driving force and slow growth?

These challenges in understanding design often come down to three issues. First, how do particular biochemical mechanisms constrain response to changed conditions? Second, how do environmental and demographic conditions alter the fitness costs and benefits of different metabolic attributes? Third, how do the mechanisms and fitness consequences combine to shape observed patterns of organismal design?

Comparative predictions provide the best approach to solving those puzzles of design. Later chapters develop comparative predictions. This chapter emphasizes observed patterns of variation and mechanistic detail, the basis for formulating hypotheses about metabolic design.

12.2 Overflow Metabolism: Mechanisms

Aerobically metabolizing cells break down glucose through glycolysis, the tricarboxylic acid cycle (TCA), and the electron transport chain. Most of the captured free energy comes from the coupling of electron transport

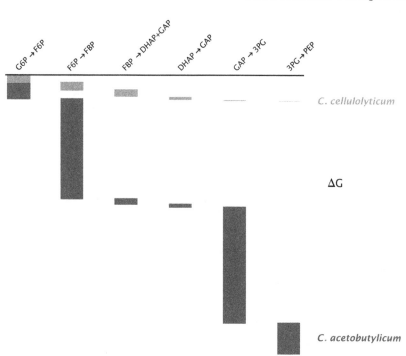

Figure 12.3   Free energy change (ΔG) between six intermediates in glycolysis
(Fig. 12.1a). The height of each rectangle shows the relative free energy change
between intermediates. The cumulative change is the sum of the changes for
intermediate components. Light gray for *Clostridium cellulolyticum* and dark
gray for *C. acetobutylicum*. For the first difference on the left, the change for
acetobutylicum combines the heights of both rectangles. Data from supplementary table 12 of Park et al.[310]

with oxidative phosphorylation to drive the ATP–ADP disequilibrium
(Fig. 12.2).

In environments with low glucose and sufficient oxygen and nutrients,
aerobically metabolizing cells process almost all of the sugar through
the full sequence. As glucose availability increases up to an intermediate
switch point, growth rate rises steadily and cells continue to catabolize
through the full sequence.

Additional glucose beyond that intermediate switch point changes
metabolic processing. Cells continue to pass some of their glycolytic
output through the TCA cycle but also excrete excess glycolytic products
such as acetate, ethanol, or lactate (Fig. 12.4).

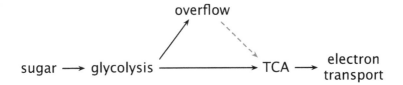

Figure 12.4 Catabolic flux and glycolytic overflow in cells capable of aerobic respiration. As sugar intake and glycolytic flux rise, cells may excrete glycolytic products. In *E. coli* at low sugar intake rate, all flux passes through the TCA cycle and electron transport, with no glycolytic overflow.[28] As intake rises, post-glycolytic flux does not keep up and excess glycolytic flux overflows as excreted acetate. In *S. cerevisiae*, rapid sugar intake associates with excreting post-glycolytic flux as ethanol. After consuming the sugar, yeast cells may shift to catabolizing the ethanol through the TCA cycle, electron transport, and oxidative phosphorylation (dashed arrow).[49]

Increased glucose uptake and faster growth rate associate with glycolytic excretion in many bacterial, yeast, and mammalian cells. The glycolytic excretion is called *overflow metabolism*.[78,326,452] Yeast studies typically label this process the *Crabtree effect* and mammalian studies the *Warburg effect*.[316]

As glucose availability rises above the switch point for overflow metabolism, growth rate continues to increase (Fig. 16.2). But cellular efficiency declines, measured as the biomass yield per gram of glucose taken up.[326]

The decline in efficiency associated with overflow metabolism likely occurs because almost all ATP production per glucose molecule happens in the post-glycolytic pathways (Fig. 12.2). Overflow metabolism excretes most of the potentially available negative entropy that drives ATP production. Oxygen and nutrients seem to be fully supplied, so it appears that cells are wasting food resources.

In the following subsections, I describe several mechanisms that may explain overflow metabolism.[78] I then discuss how the mechanistic explanations must be considered in the broader context of fitness costs and benefits. Those design forces shape metabolism subject to the constraint forces imposed by mechanism. The interplay between design and constraint forces provides the basis for comparative predictions.

THERMODYNAMIC INHIBITION REDUCES GLYCOLYTIC DRIVING FORCE

Overflow metabolism suggests that the post-glycolytic pathways hit some limit. If flux slows through those later pathways, then the glycolytic products build up in concentration.

When the concentration of a product rises to its equilibrium level, the free energy driving force decreases to zero (eqns 11.7, 11.8). Thus, excess glycolytic products extinguish the driving force of glycolysis, bringing cellular metabolism to a halt. Excretion of glycolytic products relieves this thermodynamic inhibition, increasing the glycolytic driving force.

Greater glycolytic driving force speeds the rate of glycolysis but also discards a lot of free energy, reducing the capture of free energy in the ATP–ADP disequilibrium. Overflow metabolism enhances catabolic rate and reduces yield efficiency.

In environments that favor rapid growth, the fitness benefit of maintaining rapid glycolytic ATP production may outweigh the cost of reduced efficiency. In that case, the fitness benefit from rapid ATP production provides a sufficient explanation for glycolytic overflow.[316]

Next, I summarize three mechanisms that may constrain flux downstream from glycolysis, causing thermodynamic inhibition.[78]

NADH–NAD$^+$ REDOX IMBALANCE INHIBITS FLUX

Thermodynamic inhibition in metabolism may be expressed in terms of redox potential.[467] I briefly describe redox potential. I then present the NADH–NAD$^+$ redox imbalance that may cause overflow metabolism.

In overflow metabolism, the excretion of glycolytic products suggests that flux through post-glycolytic pathways has slowed. As those lower pathways back up, cells discard some of their glycolytic products to maintain thermodynamic driving force.

From a thermodynamic perspective, what exactly is backing up and what is being discarded to bring the pathways back into thermodynamic balance? In other words, what is the proper currency to measure backup, excretion, and balance? Redox potential is the proper currency.[467]

Redox potential measures how strongly molecules attract and hold electrons. Entropy increases as electrons attach to more strongly attracting molecules. Molecules that hold electrons relatively weakly have relatively lower entropy and higher free energy. Much of the negative entropy in food arises from the relatively weakly held electrons.

Entropy increases as the electrons flow through catabolism from weakly attracting food to strongly attracting final electron acceptors, such as oxygen. Catabolism is the orderly processing of that electron flow, designed to capture food's negative entropy in molecular forms that can drive cellular processes.

Overflow metabolism arises from an imbalance in the electron flow from electron donors to electron acceptors. Electron donors are described as reducing their molecular partners because they transmit negatively charged electrons that reduce the charge of the recipient molecules. Electron acceptors oxidize (increase) the electric charge of their partners by taking away negatively charged electrons.

The flow of electrons between reduction-oxidation (redox) pairs determines much of the thermodynamic flux of entropy and free energy in metabolic reactions.

In the process of breaking down food, catabolic processing transfers some of the food's weakly held electrons to special molecules that are kept in disequilibrium, such as the transfer to ATP kept in disequilibrium against ADP. The disequilibrium of those special molecules hold for later use relatively weakly held electrons and their high potential thermodynamic driving force.

In glycolysis and the TCA cycle, some of the negative entropy in food's electrons is captured by the $NADH-NAD^+$ disequilibrium in the reaction

$$NAD^+ + H^+ + 2\,e^- \longleftrightarrow NADH.$$

An input of free energy can push the concentration of NADH above its equilibrium level. The free energy comes from coupling this reaction to a spontaneous catabolic reaction that loses free energy. The total free energy change of the coupled reactions is negative, as it must be to guarantee the increase of entropy required for all aggregate changes.

In later reactions, dissipating the $NADH-NAD^+$ disequilibrium pulls relatively weakly held electrons away from NADH toward a more attractive partner. That redox electron flow increases entropy, providing a driving force for other reactions.

The loss of electrons from NADH to more strongly attracting molecules is the primary driver of the electron transport chain, which transfers to ATP the free energy contained in the relatively weakly held electrons of NADH.

We can now return to the problem of overflow metabolism. In essence, some aspect of the TCA cycle or the electron transport chain fails to keep up with the electron flux from highly reduced food toward more strongly electron-attracting molecules.

Glycolytic products attract electrons more strongly than the initial food source. But if those glycolytic products cannot pass their electrons on to even more strongly attracting molecules in the downstream TCA cycle and electron transport chain, then the electrons at that intermediate redox level build up in those glycolytic products.

The buildup in the concentration of electrons with intermediate redox potential in glycolytic products moves the system toward its equilibrium. That reduction in disequilibrium slows the thermodynamic flux of entropy, impeding the flow of electrons through glycolysis. Cells may relieve that electron flux inhibition by excreting glycolytic products, maintaining sufficient disequilibrium.

What causes the backup in electron flux in the downstream TCA cycle or electron transport chain?

In *E. coli*, the NADH–NAD$^+$ ratio rises sharply at the onset of overflow metabolism.[423] That extreme disequilibrium means that there is little NAD$^+$ available to make NADH by accepting electrons that flow through the TCA cycle. Overflow metabolism functions to excrete excess weakly held electrons that cannot flow through a redox potential gradient in the TCA pathway. Studies of the yeast *S. cerevisiae* also show that an NADH–NAD$^+$ imbalance associates with overflow metabolism.[186,424]

To study the role of the NADH–NAD$^+$ ratio, Vemuri et al.[423] created an *E. coli* strain that overexpresses an NADH oxidase. That enzyme lowers the NADH–NAD$^+$ ratio. The lower ratio associates with more NAD$^+$ to accept electrons and produce NADH. Maintenance of the NAD$^+$ electron acceptor allows the continuous flow of electrons from glycolysis through the later pathways at higher glucose uptake rates.

However, high glucose uptake rates often trigger repression of some genes in the TCA cycle and electron transport. Reduced enzyme or cytochrome levels constrain flux, creating another barrier to flow through the post-glycolytic pathways.

To prevent repression of post-glycolytic pathways, Vemuri et al.[423] knocked out a transcription factor gene, *arcA*. The combined excess NADH oxidase and knockout of *arcA* restored electron flux balance and allowed cells to catabolize more glucose through the TCA cycle and electron transport. Enhanced catabolic processing reduced or eliminated overflow excretion of glycolytic products, even at high glucose uptake rates.

Excess NADH oxidase causes futile cycling between electron acceptance that transforms NAD^+ to NADH and electron loss that reverses the transformation. The cycling is futile because the electron flow does not drive useful biochemical transformations.[348] Associated with that futile cycling and waste of free energy, excess NADH oxidase reduces biomass yield relative to the wild type.

In summary, a rise in the $NADH–NAD^+$ ratio at high glucose uptake rates causes redox imbalance and associates with excretion of glycolytic products. The imbalance also associates with repression of the TCA and electron transport pathways.

Excess NADH oxidase and loss of the repressor for later metabolic pathways restore redox balance, enhancing flux through the full aerobic catabolic pathway. However, the engineered mutant strain has lower biomass yield per gram of glucose taken up. Reduced yield may occur because of the futile cycling of NADH and NAD^+ or the higher proteome cost of the post-glycolytic aerobic pathways (p. 168).

Vemuri et al.'s[423] experimental study shows that relieving the $NADH–NAD^+$ redox imbalance can restore post-glycolytic flux and reduce the overflow excretion of glycolytic products. In natural isolates that suffer $NADH–NAD^+$ redox imbalance at high glucose uptake rate, what mechanism causes the buildup of that redox imbalance? The following subsections consider two possibilities.

MEMBRANE SURFACE TRADEOFF BETWEEN GLUCOSE UPTAKE AND ELECTRON TRANSPORT

On *E. coli*'s inner cytoplasmic membrane, glucose transporters must compete for membrane space with electron transporters. As growth rate rises and demand on membrane proteins increases, a tradeoff may occur between glucose uptake and electron transport.[320,399,472]

At moderate metabolic rate, there is enough membrane space for cells to balance glucose uptake and the final electron transport steps of aerobic ATP production. At high metabolic rate, demand for glucose uptake may crowd out electron transport. By this hypothesis, fast-growing cells excrete excess glycolytic products because limited electron transport capacity prevents some glycolytic products from flowing through the post-glycolytic pathways.[399,472]

Normally, the electron transport chain moves the relatively weakly held electrons of NADH to oxygen, which strongly attracts electrons. That process lowers NADH concentration and raises NAD$^+$ concentration.

When electron transport capacity becomes limiting, the buildup of NADH causes redox imbalance. Excreting glycolytic products slows creation of NADH in the TCA cycle, which reduces the NADH–NAD$^+$ disequilibrium and thermodynamic inhibition.

Szenk et al.[399] studied the tradeoff between glucose uptake rate and electron transport capacity under the assumption of limited membrane space. In their theoretical analysis, abundant glucose favors allocating additional membrane space to glucose uptake, causing more glycolytic flux than can be processed by electron transport. With limited electron transport capacity, cells must excrete overflow glycolytic products to maintain catabolic flux.

According to Szenk et al.'s[399] calculations, high glucose uptake and glycolytic overflow maximize the efficiency of ATP production per unit membrane area, enhancing the rate of ATP production.

Zhuang et al.[472] suggested that eukaryotic cells may face a similar membrane limitation. Glycolysis occurs in the cytosol, yielding pyruvate. Mitochondria take up pyruvate through active transport across the inner mitochondrial membrane.[272] Mitochondria also use their inner membrane for the electron transport chain.

Limited mitochondrial membrane space may create a tradeoff between electron transport and uptake of pyruvate or other nutrients. That mitochondrial membrane limitation may cause cells to excrete excess glycolytic products at high glucose uptake rates.

PROTEOME TRADEOFF BETWEEN CATABOLISM AND BUILDING BIOMASS

The proteome is the aggregate cellular protein content. The proteome efficiency of a cellular process can be measured by the amount of protein required to drive the process.

The proteome efficiency of catabolic pathways may be expressed as ATPs produced per unit proteome. Molenaar et al.[279] suggested that glycolysis by itself is more efficient than the full catabolic pathway from sugar uptake through final aerobic processing.

In other words, glycolysis makes more ATPs per proteomic unit than does full processing through the TCA cycle and electron transport, even though glycolysis makes fewer ATPs per sugar molecule. Glycolysis is more efficient per unit proteome, whereas full aerobic processing is more efficient per unit carbon input.[284]

This theory predicts overflow metabolism. When the sugar uptake rate is high, the associated fast growth rate imposes strong proteomic demand for the proteins that aid in building biomass and replicating cells. Strong proteome demand for growth favors glycolytic catabolism, which is more proteome efficient. The switch toward glycolytic catabolism at high sugar uptake rate causes overflow excretion of glycolytic products.

When sugar uptake rate is low, the associated slow growth rate imposes weak proteomic demand. Limited carbon imposes a stronger constraint than does total protein. Cells gain by using the more carbon efficient post-glycolytic pathway of aerobic respiration.

Basan et al.[28] tested the prediction that stronger proteome limitation increases glycolytic-dominated catabolism and overflow excretion. To test this prediction, they overexpressed the LacZ protein in *E. coli.* The more LacZ, the greater the proteome limitation will be for the expression of other proteins. They observed that greater proteome limitation enhanced overflow metabolism, supporting their prediction.

The theory depends on the assumption that glycolysis is more proteome efficient for ATP production than respiration via the TCA cycle, electron transport, and oxidative phosphorylation. Basan et al.[28] used quantitative mass spectrometry and ribosome profiling to measure proteome efficiency per ATP produced. They estimated that glycolysis is approximately twice as efficient as respiration.

In summary, under high glucose uptake rate and fast growth, limited proteome capacity constrains cells. Proteome limitation favors the more proteome efficient glycolytic pathway over the less efficient respiration pathways, leading to overflow excretion of post-glycolytic products.[298]

Under low glucose uptake rate and slow growth, limited carbon availability constrains cells. Carbon limitation favors using the more carbon efficient respiration pathways rather than overflow excretion of post-glycolytic products.

12.3   Overflow Metabolism: Design Puzzles

Why do cells excrete glycolytic products, which contain most of food's potential free energy? The prior section summarized various mechanistic explanations. For example, membrane space may be limited or proteome costs may dominate. Those mechanistic explanations arise from biophysical constraints.

Constraining mechanistic forces determine what is possible and therefore play a necessary role in the study of design. But constraining forces are not sufficient to explain biological design. We must also consider what I have called the *design forces.*

Design forces determine the relative strength of fitness components, such as rate and yield. Rate is the speed at which cells reproduce. Yield is the total amount of reproduction per unit of food intake. Environmental and demographic factors influence the relative fitness value of growing fast versus growing efficiently.

Constraint forces determine the directions along which design forces can move traits. A crowded membrane surface imposes a tradeoff between the density of glucose transporters and the density of electron transporters. Increasing glucose uptake by adding more transporters often increases growth rate. But the crowding of the membrane with glucose transporters imposes reduced aerobic capacity, lowering yield.

Similarly, when demand for cytosol proteins exceeds space or resources, a proteome constraint imposes a tradeoff between enzymes for catabolism and enzymes for growth. Maximizing growth rate or balancing rate and yield may depend on this mechanistic constraint.

At first glance, the forces of constraint seem most compelling. Limited room on membranes or within the cytosol must impose essential tradeoffs. Experiments that push cells to their limits gain information about those physical constraints. Observed cellular traits may follow along the paths set by the constraints.

Do such studies of constraints solve the puzzles of design? No, for two reasons.

First, constraints are rarely fixed with regard to biological function. For example, glucose uptake may be enhanced by making the membrane more permeable.[318] Because space is not the only factor that influences uptake, constraints based solely on membrane surface area can mislead.

Membrane permeability may also affect electron transport efficiency, creating an alternative physical constraint between glucose uptake and aerobic respiration. In general, biological functions may be linked through multiple physical factors.

Experimentally, one can choose to push up against any one of the many physical factors that constrain a pair of traits. One will often see significant consequences because physical constraints impose strong forces. One finds what one looks for.

With regard to design, it is difficult to know which of the constraining forces is most important. Different circumstances push cells up against different limits. Potentially, all of them could be important. But under normal operating conditions, only a few of the limits or maybe none of them imposes strong constraining forces on design.

The second reason that studying constraints by themselves cannot solve puzzles of design is that the forces of design can tune physical constraints. For example, the mechanisms by which membrane permeability varies may allow partial decoupling of glucose uptake from electron transport efficacy. Maybe a partial decoupling of those functions requires extra cost. If it can be done, is it worth it?

The worth depends on the components of fitness affected by the different functions. And the worth depends on how the fundamental forces of design weight those different fitness components.

Maybe the fact that membrane permeability can also influence antibiotic defense is important.[318] If so, then the forces of design may have to balance the fitness components associated with sugar uptake, electron transport, and antibiotic defense, subject to various physical constraints.

How to Study Design

At this point, the puzzles of design may seem hopelessly complicated. And they are, if one tries to fit a particular explanation to a particular organism.

For example, one will never understand how the forces of design have shaped *E. coli* by studying a few alternative conditions that push the organism against biophysical constraints. Such experiments are very helpful. But they cannot solve puzzles of design.

Comparative predictions provide the best way to study the tuning of design and the interaction with constraining forces. Chapter 16 presents many comparative predictions for overflow metabolism.

The remainder of this chapter provides additional background for comparative predictions. For example, Section 12.4 discusses laboratory evolution experiments, which push against particular constraints. The evolutionary response reveals whether design follows the hypothesized constraints or other forces override those constraints.

The following subsection considers natural genetic variation, which suggests how organisms adapt to different conditions.

NATURAL VARIATION

The fission yeast *Schizosaccharomyces pombe* has the capacity for aerobic respiration. That catabolic process typically transforms sugar into carbon dioxide and water. The full cascade consumes essentially all of the usable negative entropy in the initial food input, providing a large free energy gradient to drive metabolism and physical work.

Many natural isolates of *S. pombe* limit catabolism primarily to the initial glycolytic cascade, in spite of their capacity for full aerobic respiration. The cells excrete glycolytic fermentation products, a form of overflow metabolism.

Kamrad et al.[202] studied the balance between glycolytic fermentation and post-glycolytic respiration in *S. pombe*. Among 161 isolates, 18 strains depended more strongly on respiration, whereas 143 depended primarily on glycolysis.

The rare respiration-dominant strains associated with a low activity variant of the pyruvate kinase gene. Pyruvate sits at a key branch point between glycolytic and post-glycolytic pathways (Fig. 12.2).

The laboratory strain of this species has the low activity variant and respiration-dominant metabolism. When the high activity variant was substituted into the laboratory strain, its metabolism changed to glycolytic-dominant.

A single nucleotide polymorphism determined the low and high activity kinase variants. The strains with increased pyruvate kinase activity had broad transcription and protein expression changes relative to the strains with low activity. Expression levels in the high activity strains associated with enhanced glycolysis, reduced post-glycolytic pathways of respiration, and other biochemical changes.[163]

Several other genetic variants also influenced the balance between glycolysis and respiration. That genetic variability provides wide scope for the forces of design to tune metabolic processing.

What sort of tuning might be involved? In particular, what fitness components might vary? Kamrad et al.[202] studied four fitness components: growth rate, biomass yield, survival, and stress resistance.

The 143 strains with higher pyruvate kinase activity had greater glucose uptake rate and glycolytic flux. The greater glycolytic flux associated with increased growth rate and decreased biomass yield when compared with the 18 strains with lower pyruvate kinase activity and a more respiration-dominant metabolism.

The high activity variants with greater glycolytic flux had a significantly higher survival rate during stationary phase. Survival was measured as the proportion of nondividing cells in stationary phase that divide after adding more food. After 3 days in stationary phase, 25.3% of the high glycolytic variants survived, whereas only 6.5% of the low glycolytic variants survived.

The low activity variants with greater respiration were better at tolerating oxidative stress. Respiration normally produces free radicals that impose oxidative stress.[420] Those free radicals may enhance cellular expression of mechanisms to tolerate oxidative stress.

Enhanced tolerance to internally caused oxidative stress may raise tolerance to external sources of oxidative stress. Other species or particular environments can create strong oxidative stress by producing hydrogen peroxide or other free radicals.[381]

In summary, the common variant with high activity pyruvate kinase had enhanced glycolytic flux, reduced respiration, faster growth rate, lower biomass yield, greater survival under starvation, and lower tolerance of oxidative stress.

Lab conditions are often thought to be especially favorable for high growth rate. However, the lab strain of *S. pombe* has the low activity variant of pyruvate kinase associated with relatively slower growth.

Lab conditions in this species may impose relatively strong oxidative stress, against which the lab strain's low activity variant provides better protection than the high activity variant.[202] If so, the lab strain is tuned for oxidative stress resistance rather than high growth rate. Whether true or not in this particular case, the possibility that cells tune metabolism to raise tolerance to oxidative stress suggests that assuming growth rate optimization may mislead about the forces of design.

In this example, design cannot be inferred from a single tradeoff imposed by a mechanistic constraint, such as membrane-constrained food uptake versus electron transport. One must also consider how stress resistance and other design forces change the weighting of different fitness components in response to changing environmental conditions.

To reveal the forces of design, one must make comparative predictions about how changing environmental conditions alter traits along the paths allowed by the forces of constraint.

12.4   Evolutionary Timescale

Constraints depend on evolutionary timescale. This section emphasizes laboratory evolution experiments, which analyze a short timescale. To set the context, I first discuss varying timescales in other types of study.

PHYSIOLOGICAL VARIATION

On a zero evolutionary timescale, the physiological response of a particular strain occurs in the context of a fixed genomic regulatory system. Biophysical constraints limit the possible physiological responses.

Basan et al.[28] measured the proteome constraint in *E. coli* (p. 168). As glucose uptake and glycolytic flux rise, their analysis suggested that a cellular limit on total protein favors overflow excretion of glycolytic products rather than an increase in post-glycolytic flux.

Limits on the allocation of proteins to different cellular functions may indicate a general force of constraint that acts on evolutionary design. However, in the absence of an evolutionary comparison, there is no direct evidence that a proteome constraint limits evolutionary change. During an evolutionary response, cell size may change, altering the intensity of the proteome size limit (p. 200). Or other constraints may dominate.

GENETIC VARIATION

On a short evolutionary timescale, natural genetic variation may reveal the forces of design. The evolutionary response associated with observed genetic variation typically remains confined to small changes within the current physiological system.

The prior section summarized genetic variation in natural isolates of *S. pombe*. The observed genetic variation suggested that varying oxidative stress in different habitats alters the fitness costs and benefits associated with overflow metabolism. In that case, metabolism is shaped by evolutionary forces of design imposed by the environment rather than by an internal constraint such as proteome limitation.

Taxonomic Variation

On a longer timescale, evolutionary analysis by the classic comparative method contrasts different species, genera, and higher taxonomic groups. The method associates varying environmental challenges with varying organism traits.[173]

The comparative method also emphasizes correction for shared evolutionary history. Suppose two closely related species of the same genus differ from other organisms. In that case, it is likely that those two species are similar because they inherited the same evolutionary modification rather than separately evolved that modification in response to the same evolutionary challenge.

Microbial Metabolic Design

My analysis of microbial metabolism emphasizes the shorter timescales. On shorter timescales, one has a better chance to isolate forces and partial causes of design. Longer evolutionary timescales make it increasingly difficult to correct for other causes or to assume that those other causes can be ignored.

Recent advances in lab studies, genomics, and other techniques have opened the study of shorter timescales. The great diversity of microbes and their rapid evolution provide good opportunities to match the scale over which forces act to the scale over which traits change, allowing one to isolate partial causation in the study of design.

Experimental Evolution

Laboratory evolution typically operates on a very short evolutionary timescale. In the lab, one can impose a particular environment and then observe evolutionary change in response to that environment.

The short timescale and strong selective pressure often push cells up against particular physiological constraints. The evolutionary response can provide insight into the interaction between conflicting design and constraint forces.

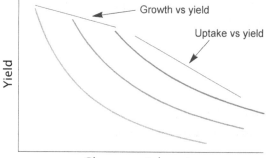

Figure 12.5   Plot of $y = \mu/q$ for yield, y, growth rate, μ, and glucose uptake rate, q. The three curves from left to right are for $\mu = 1, 1.55, 2.1$. The value of μ is constant along each curve. Physiological constraints may impose a negative tradeoff between growth rate and the maximum value of yield.

Cheng et al.'s[65] recent experimental evolution study of overflow metabolism in *E. coli* provides an interesting example. Glucose availability and other conditions remained constant. Changes in traits reflect evolutionary response to the constant experimental conditions.

Growth rate increased by approximately 50%, a very strong evolutionary response. Three other variables changed with increased growth rate: glucose uptake rate, acetate excretion rate, and biomass yield.

To interpret the evolutionary response, consider the relations between yield, growth rate, and glucose uptake rate (Fig. 10.1) as

$$y = \mu/q. \tag{12.1}$$

Here, y is yield in grams of biomass per gram of glucose, μ is growth rate in grams of biomass per unit time, and q is the glucose uptake rate in grams of glucose per unit time. This expression follows from the definitions of the terms and does not require any particular assumptions.

For a given growth rate, μ, the other two variables have the inverse relation shown in Fig. 12.5. The value of μ is the same along each curve. An increase in μ shifts a curve to the right. The negative tradeoff between uptake rate, q, and yield, y, arises from $y \propto 1/q$ in eqn 12.1.

In Cheng et al.'s[65] study, all independently evolved lines significantly increased their growth rates (Fig. 12.6). For similar growth rates, the evolved lines varied widely in their biomass yield (y) and glucose uptake (q) traits. The variability in those two traits followed along the tradeoff

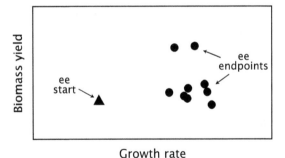

Figure 12.6 Experimental evolution increases growth rate, with biomass yield changing in an apparently neutral and uncorrelated way. When the experimental evolution (ee) starting strain was subjected to natural selection favoring faster growth, the nine independently evolved lines changed to the endpoints. Redrawn from Fig. 1A of Cheng et al.[65]

curves $y \propto 1/q$ in Fig. 12.5 imposed by the definition of the variables (fig. 3A of Cheng et al.[65]).

In the evolved lines, why do the uptake rate and yield vary so widely for similar growth rates? In these experiments, selection was imposed in a way that limits time for growth rather than limits sugar for growth. Because the imposed design force pushes strongly on biomass produced per unit time (growth rate) and weakly on biomass produced per unit sugar (yield), the yield is effectively a neutral trait.

Under these conditions, an evolutionary change that increases growth rate is favored independently of its consequences for yield. Thus, the various evolved lineages tend to explore alternative physiological mechanisms to achieve the same level of increased growth.

In terms of the evolutionary forces of design, the interesting problem concerns how the environment imposes particular fitness costs and benefits. For example, an experiment that imposed selection on both growth rate and yield would, in theory, favor phenotypes that followed along the upper growth rate versus yield tradeoff line in Fig. 12.5. One could alter the selection intensity on the rate and yield components, potentially favoring movement of the evolved lineages along the rate-yield tradeoff line.

In terms of the forces of constraint, the interesting problem concerns how physiological and biophysical mechanisms impose limits on particular variables and associations between those variables. For example, what mechanisms allow cells to evolve higher growth rates? What mechanisms relate the observed variations in glucose uptake rate to variations in yield for a given growth rate?

This study did not directly measure mechanistic aspects associated with the observed evolutionary changes. Instead, the authors used modeling approaches to analyze plausible mechanistic explanations. For the observed variation in glucose uptake rate and yield, their metabolic modeling suggested a possible role for an NADH–NAD⁺ imbalance, discussed earlier on p. 164.

By that mechanism, the more that cells process post-glycolytic products through the TCA cycle and the electron transport chain, the lower the glucose uptake rate and the higher the yield for a particular growth rate will be. Lower uptake rate also associates with lower post-glycolytic excretion of acetate.

To increase post-glycolytic flux through the TCA cycle and electron transport, cells must relieve the NADH–NAD⁺ imbalance that builds at high growth rate. That imbalance may build because electron transport cannot dissipate the NADH–NAD⁺ disequilibrium fast enough to offset the increasing imbalance produced by the TCA cycle (p. 164).

The metabolic models by Cheng et al.[65] suggest that cells may relieve the NADH–NAD⁺ imbalance by switching to alternative, faster, and less efficient electron transport components. For example, Zhuang et al.[472] note that three different cytochromes used in electron transport trade off speed versus efficiency. By adjusting the ratios of those cytochromes, cells could adjust the speed versus efficiency of electron transport.

Less efficient electron transport captures less free energy in the ATP–ADP disequilibrium, dissipates the NADH–NAD⁺ disequilibrium faster, and increases flux through electron transport. That increased flux through aerobic respiration lowers acetate excretion, raises yield, and lowers the glucose uptake associated with a particular growth rate.

How should we interpret these details about evolutionary response and physiological mechanisms? I prefer to synthesize the existing facts and current theories into a series of comparative predictions. Those predictions emphasize how environmental changes alter the evolutionary forces of design and the associated weighting of the various fitness components. Those evolutionary forces can tune traits via the particular physiological mechanisms that impose forces of constraint.

Of course, we may misinterpret facts and develop incorrect theories. Then the comparative predictions should fail, exposing the problem.

I develop comparative predictions in Chapter 16. I finish here with one brief prediction. As experimental environments impose stronger selection on improved yield, the evolved lines should converge toward more efficient metabolic pathways associated with lower glucose uptake rate. For example, we may expect a tuning in electron transport, NADH–NAD$^+$ imbalance, or other pathways that enhance yield efficiency.

A couple of tentative conclusions follow. First, physiological mechanisms provide much insight into design. But mechanistic constraints can rarely explain design. For example, constraints play different roles in environments that favor high growth independently of yield versus environments that strongly select on yield efficiency.

Second, one can study the forces of design without knowing about physiologically imposed forces of constraint. Environments that favor growth more than yield will tend to produce different traits when compared with environments that favor yield more than growth. That comparative prediction does not depend on mechanism.

Sometimes the multiple overlapping tradeoffs and the complexity of the underlying physiology imply that it is best to start with broad comparative predictions based on the forces of design, ignoring constraints imposed by mechanism.

Ultimately, including constraints and potential evolutionary changes in mechanism will provide more precise predictions and greater insight into design. Chapter 16 develops many comparative predictions.

12.5 Alternative Glycolytic Pathways

Prior sections considered glycolytic design in terms of free energy driving force, proteomic efficiency, ATP production, and membrane permeability. Those attributes affect growth rate and biomass yield per gram of sugar input. Glycolytic and membrane attributes also affect oxidative stress and sensitivity to antibiotics.

This section compares alternative glycolytic pathways.[208] The different pathways reflect the forces that shape alternative metabolic designs.

REDOX GRADIENTS AND METABOLIC PRODUCTS

Glycolysis begins by taking up glucose or changing molecules into forms that can be passed into glycolysis. The initial food molecules typically hold electrons relatively weakly. Catabolic processing transfers those weakly held electrons to stronger electron acceptors through a gradient of redox reactions.

The cascade of catabolic steps captures some of food's negative entropy into various storage systems. Each storage system is a chemical disequilibrium that can be used to drive other reactions. I mention a primary function for each of the three major systems.

The NADH-NAD$^+$ disequilibrium drives the ATP-ADP disequilibrium through oxidative phosphorylation (Fig. 12.2). The ATP-ADP disequilibrium powers much of cellular work, including biosynthesis, motive force, and active transport.[86] The NADPH-NADP$^+$ disequilibrium drives the building of complex molecules and the mitigation of oxidative stress.[394,428]

We can describe the alternative glycolytic pathways by their negative entropy capture in the three storage systems. We can also describe the pathways by their proteomic cost for catalytic enzymes.

ALTERNATIVE PATHWAYS

The Embden–Meyerhof–Parnas (EMP) pathway occurs in all domains of life. From glucose to pyruvate, the pathway drives the following two reactions to increase the NADH-NAD$^+$ and ATP-ADP disequilibria,[428]

$$2\,NAD^+ + 2\,H^+ + 4\,e^- \longrightarrow 2\,NADH$$
$$2\,ADP + 2\,Pi \longrightarrow 2\,ATP.$$

We say that the pathway produces 2 ATP and 2 NADH per glucose molecule to describe the increased disequilibria.

The Entner-Doudoroff (ED) pathway occurs in many Bacteria and some Archaea.[48,108,208] Among Eukarya the pathway has been described in a few plants, apparently obtained by gene transfer from cyanobacterial ancestors of plastids.[64] The ED pathway yields 1 NADH, 1 ATP, and 1 NADPH.

The hexose monophosphate pathway (HMP), also known as the pentose phosphate pathway (PPP), occurs widely across the Bacteria and Eukarya.[394] Some Archaea contain parts of the HMP pathway and may use other reactions to achieve similar function.

The HMP pathway produces 2 NADPH and also provides molecular precursors for nucleic acids and some amino acids. In most organisms, this pathway drives maintenance, growth, and protection against oxidative stress.[394] Pentoses such as xylose, arabinose, and ribose may enter glycolysis through this pathway.[208]

In summary, the typical disequilibrium productivities for the three major pathways are

$$\text{EMP} \longrightarrow 2\,\text{NADH} + 2\,\text{ATP}$$
$$\text{ED} \longrightarrow 1\,\text{NADH} + 1\,\text{ATP} + 1\,\text{NADPH} \qquad (12.2)$$
$$\text{HMP} \longrightarrow 2\,\text{NADPH}.$$

The biochemistry and yield of these pathways vary across prokaryotes.[26,48,208]

KEY ATTRIBUTES AND FITNESS CONSEQUENCES

This subsection links glycolytic pathways to five components of cellular performance and fitness. The following subsection outlines associated puzzles of metabolic design.

1. Yield.—Equation 12.2 summarizes the standard storage disequilibrium yields for the alternative glycolytic pathways. Cost-benefit analyses in the literature typically consider only ATP.[108] However, NADH and NADPH disequilibria also provide benefit through alternative functions.

Different environmental challenges alter the benefits associated with the different disequilibria. How do microbes tune their usage of alternative glycolytic pathways and disequilibria in response to varying environmental challenges? The next subsection considers that question.

2. Driving force and rate.—We can partition into components the total free energy change of coupled reactions in a pathway,

$$\Delta G = \Delta G_p - \Delta G_y.$$

The total change, ΔG, is the maximum potential change, ΔG_p, when the pathway does not drive coupled reactions that decrease entropy. The potential change must be discounted by ΔG_y, the free energy yield of entropy-decreasing coupled reactions. In glycolysis, the storage disequilibria hold the free energy yield.

The ratio $\Delta G_y/\Delta G_p$ describes the yield capture efficiency, the fraction of a pathway's potential free energy change captured by the storage disequilibria. Four consequences follow.

First, greater yield efficiency reduces the overall driving force, ΔG.

Second, lower driving force reduces the pathway flux rate when holding constant the reaction resistance, caused by factors such as enzyme concentrations. Thus, the yield per input molecule trades off against the rate at which the pathway processes input molecules.

Third, the common method of counting the number of ATP, NADH, and NADPH molecules produced may not accurately reflect the driving force and yield. For example, if the NADH–NAD$^+$ disequilibrium is relatively high, then converting another NAD$^+$ to NADH consumes more free energy than if the NADH–NAD$^+$ disequilibrium is relatively low. Similarly, the potential free energy change through the pathway, ΔG_p, depends on the concentrations of the molecules along the reaction sequence.

Fourth, metabolic efficiency can be described as the increase in the storage disequilibria per input molecule (yield efficiency) or the increase in the disequilibria per unit time (rate efficiency).

3. Proteome cost.—Another efficiency aspect concerns the amount of enzyme used to catalyze a pathway. Greater enzyme concentrations lower the resistance of reactions and increase the pathway flux rate. Enzymes impose a protein production cost.

If proteome size is limited, then allocating more protein to glycolytic throughput reduces the amount of protein that can be allocated to other functions. We may consider the yield efficiency per unit proteome cost and the rate efficiency per unit proteome cost.

4. NADPH relieves oxidative stress.—I first review redox biochemistry and oxidative stress. I then turn to the key role of NADPH and its production in the ED and HMP glycolytic pathways.

The driving force for cellular biochemistry comes from moving the weakly held electrons in food to strong electron attractors. The movement of electrons to strong attractors is called oxidation. Oxygen is a common electron attractor in oxidation, but other electron attractors also oxidize molecules.

The flow of electrons toward strong attractors in catabolism increases entropy. Cells couple that increase in entropy to other processes that lower entropy, such as driving storage disequilibria, building organic molecules, or doing physical work. The coupled processes can happen as long as the total entropy increases.

When building molecules or driving storage disequilibria for later use, reactions that enhance negative entropy often move electrons back toward molecules that hold them relatively weakly. Adding electrons to molecules is called *reduction* because more negatively charged electrons reduce overall charge.

Organic molecules tend to be in a reduced state relative to commonly encountered electron attractors that would oxidize them. Unwanted oxidizing agents impose oxidative stress by pulling electrons away from reduced molecules, which causes damage.[396]

Oxidation causes a fundamental tension of biochemistry. Moving weakly held electrons from food molecules toward strong electron attractors by oxidation provides the catabolic driving force of life. Moving weakly held electrons from useful molecules to unwanted oxidizing agents destroys life's ordered molecules.

Cells carefully control oxidizing processes in catabolism. However, unwanted oxidizers are common and very damaging. Cells have antioxidants that can counter oxidative stress.[57,396]

Oxidative stress often arises from molecules with unpaired electrons, which are highly reactive. The hydroxyl radical, $^\bullet OH$, and the superoxide anion, $^\bullet O_2^-$, are common free radicals in biology. Many other free radicals occur. Free radicals arise spontaneously as products of chemical reactions within cells. Free radicals in the environment enter the cell through the membrane.

Cells use a variety of antioxidants to control free radicals. Antioxidant processes tend to be reducing, that is, they tend to push electrons toward other molecules to counteract electron-attracting oxidizers. Thus, antioxidation requires a reducing driving force.

Cells often use an NADPH–NADP$^+$ disequilibrium to drive antioxidant reducing processes because the reaction

$$NADPH \longrightarrow NADP^+ + H^+ + 2\,e^-$$

can be coupled with other reactions to push electrons toward molecules.[57] Thus, cells must drive the NADPH–NADP$^+$ disequilibrium to maintain a store of reducing power for protecting against oxidative stress.[385]

Returning to glycolysis, the ED glycolytic pathway produces NADPH, whereas the standard EMP pathway does not (eqn 12.2). Oxidative stress may favor the antioxidant power of the ED glycolytic pathway over the EMP pathway.[63]

Electron transport in oxidative phosphorylation creates free radicals and a strong potential for oxidative damage.[42,221] Thus, greater aerobic respiratory flux through electron transport may favor ED over EMP. Alternatively, cells may process a portion of glycolytic flux through HMP/PPP to build the NADPH–NADP$^+$ disequilibrium.

5. *Membrane transport.*—Glycolytic flux depends on food uptake. Faster uptake may associate with greater transport of external oxidative factors into the cell, increasing oxidative stress and favoring ED over EMP.

Two studies provide circumstantial evidence that uptake of sugar may influence the uptake of oxidizing agents.

First, greater nutrient transport alters the uptake rate for various antibiotics.[318] Some antibiotics damage cells by oxidation.

Second, particular outer membrane pores of *Salmonella* lowered their permeability in response to external oxidizing agents.[177] Mutants that cannot reduce permeability suffered greater oxidative damage.

Thus, altered nutrient transport may influence membrane permeability, affecting uptake of other molecules and sensitivity to external oxidative stress.

If greater nutrient transport increases sensitivity to external oxidative challenge, then greater glycolytic flux may favor ED over EMP to protect against external oxidative stress.

PUZZLES OF DESIGN

What factors favor one glycolytic pathway over another?

1. *Differences in benefits and costs.*—The pathways differ in their yields for the three storage disequilibria (eqn 12.2). Each disequilibrium provides a different benefit. Matching disequilibrium production with the associated demand and dissipation rate is likely to be a key challenge.

Greater disequilibrium yield reduces the net free energy change. All else equal, greater yield lowers the driving force and the flux rate.

Oxidative stress favors NADPH production because of its antioxidant properties. Oxidative stress rises with electron transport flux, external oxidative challenge, and greater membrane permeability associated with faster nutrient uptake.

The pathways likely differ in matching glycolytic output to demand from the TCA cycle and aerobic respiration. Similarly, the pathways likely differ in excreting post-glycolytic overflow products.

Excreted glycolytic products can influence growth of neighboring cells, which may be competitive or cooperative.

The pathways provide different anabolic precursors for building organic molecules.

The pathways accept different alternative nutrient molecules and initiating reactions. For example, the HMP pathway takes pentose as a nutrient input.

Proteome cost arises primarily from producing the enzymes to catalyze reactions. More pathway reaction steps increase proteome cost. Lower driving force requires more enzymes and higher proteome cost to achieve the same flux rate.

2. EMP or ED versus HMP.—Kim & Gadd[208] state that in *E. coli* grown on glucose "about 72 percent of the substrate is metabolized through the EMP pathway, and the HMP pathway consumes the remaining 28 percent." HMP may be used for its NADPH–NADP$^+$ disequilibrium or for particular precursors needed for biosynthesis.

The split between EMP and HMP almost certainly varies with growth conditions. How do cells adjust the split between pathways in response to changes in the environment? Which of the many differing costs and benefits dominate under different conditions? How do the forces of design tune the adjustment? How do species vary in this split or in a split between ED and HMP?

Kim & Gadd[208] (their section 4.1.3) list alternative steps in the EMP pathway in different bacteria. For example, the standard pathway uses NAD$^+$-dependent glyceraldehyde-3-phosphate dehydrogenase (GAPDH). *Streptococcus bovis* and other species have an NADP$^+$-dependent variant of GAPDH, which produces NADPH. EMP production of NADPH may be important in species that lack the HMP pathway or other mechanisms to generate NADPH.

3. EMP versus ED.—Flamholz et al.[108] inferred from genome sequences whether a species has EMP or ED. In over 500 genomes of Bacteria and Archaea, most species have EMP only, relatively few species have ED only, and some species have both. Aerobes have ED significantly more often than anaerobes (Fig. 12.7).

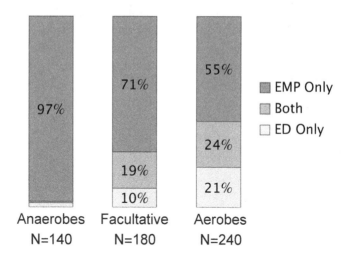

Figure 12.7 Distribution of EMP and ED glycolytic pathways among Bacteria and Archaea. Presence of a pathway was determined by analyzing genomes for the genes in that pathway. Redrawn from Fig. 6 of Flamholz et al.[108] It would be interesting to refine the analysis with phylogenetic comparative methods.[173]

EMP yields two ATP molecules per glucose and ED yields one (eqn 12.2). Thus, EMP has the greater ATP yield per sugar molecule. Anaerobes often rely on glycolysis for ATP production. Flamholz et al.[108] suggest that the strong bias of anaerobes for EMP arises from that pathway's superior ATP yield per unit carbon input.

For aerobes, most ATP comes from post-glycolytic pathways. Thus, aerobes weight the benefit of ATP yield from glycolysis less than do anaerobes. Instead, aerobes may be more strongly influenced by the cost of running the glycolytic pathway.

Flamholz et al.[108] suggested that EMP has a higher proteome cost than ED to produce the same post-glycolytic flux. Because EMP has a higher yield than ED, it likely has a lower thermodynamic driving force. The lower the driving force, the more enzyme required to catalyze flux.

Because EMP demands more protein to catalyze flux, the net ATP yield per unit proteome cost may be lower in EMP than in ED. Aerobes may sometimes be more strongly limited by the protein cost of driving glycolysis than by the limited ATP yield of glycolysis because most ATP comes from post-glycolytic pathways. Thus, according to Flamholz et al.,[108] ED is more prevalent in aerobes than anaerobes.

Alternatively, Chavarría et al.[63] suggest that ED may be favored over EMP in bacteria "to gear their aerobic metabolism to endure oxidative-related insults." They show that the greater NADPH production of ED provides the aerobe *Pseudomonas putida* with better tolerance to oxidative challenge by diamide and hydrogen peroxide.

Aerobic prokaryotes may face greater oxidative stress than anaerobes because electron transport produces free radicals.[42,221] Excess oxidative challenge for aerobes may partly explain their use of ED over EMP.

Both proteome efficiency and oxidative tolerance provide reasonable hypotheses. One could probably come up with other plausible alternatives to fit the observed pattern.[256] However, only comparative hypotheses and tests can reveal the forces of design and the forces of constraint.

4. ED lost from eukaryotes except in some plants.—The environmental challenges and natural history of aerobic yeast are similar to many aerobic prokaryotes. Why do aerobic yeast lack the ED pathway, whereas aerobic prokaryotes often have that pathway?

Perhaps the presence of mitochondria in eukaryotes explains the difference. Mitochondria confine the free radicals produced by aerobic metabolism. Processes that detoxify mitochondrial oxidative stress are separated from the cytosol, where glycolysis occurs. Thus, the antioxidant benefits of the ED pathway may provide more value to prokaryotes than they would to eukaryotes.

Currently, all eukaryotes are thought to lack the ED pathway except some plants.[64] If mitochondria and the associated processes of intracellular oxidative stress and tolerance explain the loss of the ED pathway in most eukaryotes, then why do some plants have that pathway?

Chen et al.[64] suggest that plants are typically not carbon limited because they produce their own carbohydrates by photosynthesis. Thus, greater proteomic efficiency may be more important for some plants than efficiency in ATP yield per food input, which could favor the ED pathway over the EMP pathway.[108]

Alternatively, some plants may suffer particularly strong oxidative stress[77] that requires the additional antioxidant power of the ED pathway. The puzzle remains unsolved.

In summary, the different pathways influence several costs and benefits. A focus only on yield maximization, or proteome cost minimization, or another particular dimension may miss important factors. As always, it is difficult to explain any particular aspect of design by itself because many forces may be acting simultaneously.

Comparative hypotheses focus on how a changed environment alters particular costs and benefits. Comparison isolates partial causation and reduces interference by other simultaneously acting forces.

Before I develop broad comparative hypotheses about metabolic design, it is helpful to review additional facts of metabolism. The next chapter turns to the modulation of resistance in metabolic reactions.

13 Flux Modulation: Resistance

Reaction rate depends on the thermodynamic driving force that pushes reactants toward products and on the resistance that opposes the reaction. Equation 11.9 expresses reaction flux by an approximate analogy with Ohm's law as flux = force / resistance.

Chapter 12 focused on the driving force of reactions, which is the increase in entropy between reactants and products (eqn 11.4). This chapter considers resistance, which impedes reactions.

The first section reviews how resistance alters chemical reaction flux. The second section describes mechanisms that modulate resistance. Constraints limit the control of resistance and flux.

The third section raises genetic drift as a fundamental constraining force on design. In small populations, stochasticity in reproduction can overwhelm any fitness differences between alternative traits. Drift may be particularly important when analyzing the design of metabolic control for individual biochemical reactions.

The fourth section highlights general challenges in the study of metabolic control. The fifth section lists specific problems of reaction flux. The final section notes gaps in current understanding and prospects for further work.

The design of regulatory control provides a natural extension to this book, setting a primary task for the future.

13.1   Resistance Impedes Flux

ACTIVATION ENERGY BARRIER

Organisms inevitably decompose into water, carbon dioxide, nitrogen, and other simple inorganic compounds. Those compounds have much lower free energy than the organic compounds of life. The driving force toward decay is strong. Yet decay happens slowly enough to allow life.

In other words, thermodynamic driving force tells us where things end up in the long-run equilibrium. But the thermodynamics of decay does not by itself explain chemical kinetics on the timescales that matter for life.

We must also consider the driving force of negative entropy from food, flowing from the sun or from geochemical disequilibria. And we must consider the resistance that opposes reactions.

With regard to resistance, consider that both diamond and graphite are composed of carbon atoms bound to one another. Graphite has lower free energy, so the thermodynamic driving force favors diamond to decompose into graphite. However, that happens so slowly in the conditions in which we live that essentially it does not happen. The resistance is very high.

Resistance occurs because the transition requires diamond to break its strong carbon-carbon bonds before reforming a different pattern of carbon bonding in graphite. The intermediate stage with broken carbon bonds has much higher free energy than the diamond crystal. So the intermediate transition almost never forms spontaneously, and the transformation to graphite almost never occurs.

In general, an intermediate reaction state with higher free energy than the initial reactants is called an *activation energy barrier*. Figure 13.1 shows an example. The heights of the molecular forms correspond to relative free energy levels.

The change from reactants to products, R $\longrightarrow$ P, decreases free energy and has a strong thermodynamic driving force. However, the reaction happens very slowly. The reaction must go through an intermediate, R $\longrightarrow$ C $\longrightarrow$ P, in which thermodynamic driving force works against the formation of the intermediate complex, C. Thus, C imposes high resistance, impeding the driving force that favors R $\longrightarrow$ P flux.

Mechanisms that alter driving force and resistance include changes in the concentrations of reactants and products, changes in conditions such as pH or the surrounding solvation environment, barriers to diffusion that resist encounters between reactants, catalysts that alter the intermediate complex, and so on.

The variety of biophysical mechanisms can only roughly be divided into abstract driving force and resistance categories. For example, reduced diffusion alters molecular concentrations and the thermodynamic driving force. Alternatively, one may consider limited diffusion as a physical resistance mechanism that impedes molecular motion and reaction. Nonetheless, the abstractions of force and resistance provide insight.

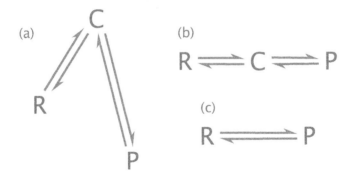

Figure 13.1   The activation energy barrier impedes a reaction with strong thermodynamic driving force. (a) The initial reactant, R, must change into an intermediate reaction complex, C, before completing the reaction by forming the product, P. Greater height of a molecular form signifies higher free energy. (b) Symbolic expression of reaction fluxes between states. We may denote each rate of transition as k_{ij}, the transition from i to j. From the left panel, we can derive the relative rates as $k_{pc} < k_{rc} < k_{cr} < k_{cp}$. (c) When the intermediate complex, C, decays rapidly and is not observed, one instead observes forward and backward transitions between the reactants and products at rates $k_{pr} < k_{rp}$. Those rates depend on the free energy driving force between the reactants and products and on the resistance imposed by the increased free energy of the intermediate complex.

Short-Term Kinetic versus Long-Term Thermodynamic Control

Useful organic molecules must be sufficiently stable so that they do not decay too rapidly. And those molecules must be sufficiently reactive so that they can be changed into other forms or be destroyed as needed.

Figure 13.2 shows the tension between thermodynamically driven decay and kinetically driven transformation between molecular forms. We start with some reactant molecules, R. Very strong driving force favors decay to inorganic products, P. But strong resistance greatly slows decay because the high free energy intermediate complex, C, rarely forms.

The initial reactants, R, can also change to an alternative molecular state, S, that may be a useful variant form. The alternative state, S, has only slightly lower free energy and relatively low driving force toward formation. However, the resistance for R $\longrightarrow$ S is relatively low, causing that transformation to happen relatively quickly.

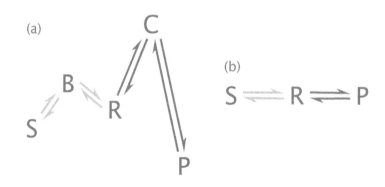

Figure 13.2   Kinetic versus thermodynamic control. Initial reactants, R, may decay to P or be modified to an alternative molecular form, S. In the short term, kinetics favors transformation to S because the lower activation barrier for B makes R ⟶ S relatively rapid. In the long term, thermodynamics favors the transformation of molecules to P because molecular forms eventually move to the lowest free energy state. (b) If the intermediates B and C transform into other molecules very quickly, one only observes the fluxes between the more stable molecular forms, represented by this diagram. See Mallory et al.[260] for flux analysis in relation to reaction barriers and thermodynamic driving force.

Figure 13.3 shows an example of the reaction dynamics. Initially, all molecules are in state R. The alternative S form arises quickly (light curve) because of the low resistance for the R ⟶ S transformation. Over a long period of time, all molecules move toward the lowest free energy state, P, because thermodynamic driving force eventually dominates (dark curve).

For cells to function, their high free energy organic molecules must not decay rapidly to low free energy inorganic components. In other words, the building, maintenance, error correction, and recycling of organic molecules requires sufficient resistance to impede the long-term driving force toward decay.

Given sufficient resistance to long-term thermodynamic decay in the R ⟶ P transition of Fig. 13.2, cells still have the short-term problem of controlling the kinetics of the R ⟶ S reaction. Altering short-term resistance provides one mechanism to modulate flux.

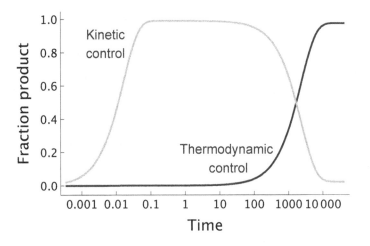

Figure 13.3 Kinetics dominates in the short term and thermodynamics dominates in the long term. The plot shows an example of the dynamics implied by Fig. 13.2. Initially, all molecules are in state R. The light and dark curves show the fraction of molecules in states S and P, respectively. The dynamics are based on linear transition rates $k_{rs} = 60$, $k_{sr} = 0.5$, $k_{rp} = 0.05$, and $k_{pr} = 0.00001$.

13.2 Mechanisms to Alter Resistance and Flux

ENZYME CONCENTRATION

A protein catalyst (enzyme) may reduce the free energy of the intermediate complex. Reduced activation energy lowers the resistance barrier between reactants and products, increasing the reaction flux.

Changing the enzyme concentration provides a mechanism to control flux. The concentration of a particular enzyme depends on four rate processes.[174] Transcription creates mRNA. Transcript decay removes mRNA. Translation creates proteins in proportion to mRNA abundance. Protein decay removes proteins.

Suppose decay rates are constant. Then enzyme concentration varies with the transcription and translation rates. Cells can make the same concentration by raising one of the rates and lowering the other. In theory, fast transcription and slow translation produce approximately the same concentration as slow transcription and fast translation.

In a comprehensive dataset, the combination of high transcription and low translation rates rarely occurred.[174]

What might favor relatively low transcription and high translation rates? Perhaps the benefits of limiting expensive transcription outweigh the costs of greater noise with low mRNA transcript numbers.

On the cost side, lower transcription rate increases noise. Noise increases because the number of mRNA transcripts may drop to the point at which stochastic production of one more or one less transcript strongly influences output.[271,293]

On the benefit side, lower transcription rate increases observed growth. Additional mRNA may be inefficient and costly for growth because each transcript is used less often to make a given amount of protein.[145,200,271]

The rarity of high transcription and low translation suggests that the loss in growth rate from excess transcription typically outweighs the gain from suppressing noise.[174]

COMPARATIVE PREDICTIONS

The tentative conclusions about the biophysical constraining forces of production cost and noise pose a puzzle about the control of enzyme concentration. How do the forces of design, which weight noise and growth components of fitness differently in different environments, affect the noise-growth tradeoff in transcription and translation? Comparatively, can we predict how changes in the environment alter expression?

Comparative tests could be applied to whole genomes, analyzing the overall tradeoff in metabolism between noise and efficiency. Or the tests could be applied to particular reactions, analyzing the noise-efficiency tradeoff in the control of particular metabolic steps.

The latter tests link flux control for particular aspects of metabolism to the ways in which changed environments alter fitness components. Natural history shapes biochemistry, the primary theme of this book.

ENZYME MODIFICATION

Modifying an enzyme can significantly change its catalytic properties. For example, adding a phosphate group to a single amino acid can alter how the enzyme binds to its substrates. Phosphorylation commonly occurs by taking a phosphate group from ATP and adding that phosphate group to an enzyme,

$$ATP + E \xrightarrow{\text{kinase}} ADP + E-Pi,$$

in which ATP gives up a phosphate group to produce ADP, and the phosphate group is attached to the enzyme as E–Pi. A kinase enzyme catalyzes the phosphorylation reaction.[428]

A phosphatase enzyme removes the phosphate group, reversing the phosphorylation modification,

$$H_2O + E–Pi \xrightarrow{\text{phosphatase}} Pi + E.$$

An enzyme can have many hundreds of amino acids. Phosphorylation or similar covalent modification typically changes only one small molecular group on one amino acid. Altering flux by enzyme modification is much faster and less costly than altering enzyme concentration through protein production and degradation pathways.

Eukaryotes rapidly adjust enzymes by kinase-phosphatase pairs or by other modifications.[189] Several studies suggest that prokaryotes also regulate flux via enzyme modification.[56,158,232]

Enzyme modification gains the benefits of speed and relatively low cost. However, small modifications typically cannot provide the level of enzymatic specificity and efficiency achieved by a custom enzyme for a specific task. Different challenges acting over different timescales likely favor different mechanisms for controlling reaction flux.[100]

ALLOSTERIC CONTROL

Small molecules can bind to enzymes, changing enzyme structure. Such allosteric change in structure often modifies the catalytic activity of the bound enzyme.[287,412]

The aggregate catalytic activity of a target enzyme can be modulated by altering the concentrations of small allosteric effectors. Allostery provides another relatively fast and inexpensive way to control flux.[319]

Aspartate transcarbamoylase (ACTase) provides a classic example of allostery.[428] This enzyme catalyzes a key step in pyrimidine synthesis and the production of nucleotides.

A later step in the pyrimidine synthesis pathway makes cytidine triphosphate (CTP). Binding of CTP to ACTase reduces enzyme activity. This allosteric binding of CTP to ACTase creates a negative feedback loop that prevents overproduction.

A separate nucleotide synthesis pathway makes purines. The cell requires a balance between purines and pyrimidines. Balance arises by an excess of purines stimulating pyrimidine production.

Cells use the purine ATP to achieve positive feedback from purines to pyrimidines. Binding of ATP to ACTase stimulates enzyme activity, creating a positive feedback that balances purines and pyrimidines.

These two allosteric modifications of ACTase create a positive and negative feedback pair that regulates nucleotide synthesis.

TRANSCRIPTION FACTORS AND PHYSIOLOGICAL RESPONSE

A typical transcription factor is a protein that binds to DNA, altering the rate of transcription and protein production of nearby genes.[24,180] A single transcription factor may bind to multiple DNA sites, altering expression for a set of genes.

Transcription factors regulate the production of enzymes that modulate the resistance and flux of reactions. Cells often initiate wide changes in their biochemical flux pathways by altering the abundance of particular transcription factors.

Shifts in enzyme concentrations typically happen on a slower timescale than covalent modification or allosteric binding of enzymes.

Space and production costs may limit the total amount of protein, including transcription factors. Suboptimal control of gene expression may occur widely in prokaryotes because of limitations on the abundance of transcription factors.[329]

Much research focuses on the biophysical structure of transcription factors and the mechanisms by which they control the production of proteins.[315] Broader design puzzles have received relatively little attention. For example, how do the life history forces of design alter the characteristics and expression of particular transcription factors?

PROTEOME LIMITATION

Space and resource constraints limit total protein production. Proteome limitation implies tradeoffs in flux because cells cannot make enough enzyme to control the flux of all reactions.[28,165,454]

Fast growing cells may reduce proteome limitation by increasing ribosome count and protein production.[343] Fast growing cells sometimes have larger cell size,[417,445] which may reduce proteome space constraints.

The potential for cells to modify constraints raises questions about design. What environmental conditions favor cells to increase proteome size? How do demographic factors influence the relative weighting of growth and other fitness components, setting the costs and benefits of cell and proteome size?

SPATIAL SEPARATION

Within the cellular cytosol, many reactions appear to be limited by diffusion.[355,359] In diffusion-limited reactions, chemical transformation from the collision of reactants in one spatial location happens faster than the time it takes for other potential reactants to re-equilibrate into spatially homogeneous concentrations.

Uneven spatial distribution of reactants impedes reaction flux, increasing resistance against the potential driving force. Creating localized reaction centers and modulating diffusion within the cytosol provide mechanisms to alter the resistance that impedes reactions.

Eukaryotes, with their internal membranes and phase-separated partitions, have greater intracellular barriers than do prokaryotes.[377] Within prokaryotes, much diversity likely occurs in the mechanisms by which reactions are localized, reactants are separated, and gradients are modulated and exploited.[109,355,398] The study of separation mechanisms is currently an active and controversial topic.[233]

MEMBRANES AND DISEQUILIBRIUM

Membranes create a primary physical barrier that impedes reactions. Cells modulate spatial gradients across membranes by altering diffusion or transport.[35,86]

Changing the concentrations across membranes modifies the resistance against reactions. Dissipating disequilibrium across membranes can drive coupled reactions that would otherwise be unfavorable. Exploiting flux across membranes to drive other processes provides a primary force for much of life.[86,467]

Chemical gradients between cells also create resistance and disequilibria. Groups of cells may exploit intercellular resistance and flux to create physiological processes across a social network.[35]

TEMPERATURE AND THERMODYNAMIC INHIBITION IN METABOLISM

Temperature affects reactions.[10] Heat speeds things up, alters entropy changes and driving force, and modifies resistance via diffusion. Temperature also influences the enzymes and regulatory mechanisms that control flux.[327]

When temperature changes, flux may be perturbed, potentially leading to product inhibition or otherwise creating bottlenecks. Temperature may more strongly perturb reactions with low net free energy change because small perturbations significantly alter the flux of those intrinsically slow reactions.

Organisms that live in stable temperatures may be tuned differently from organisms that face fluctuating temperatures. With fluctuation, organisms may require special designs to cope flexibly with altered rates. They may also need special functions to clear product inhibition or other bottlenecks that arise from concentration mismatches between flux pathways.

Reactions typically occur as parts of pathways. Thus, temperature effects must be considered in the context of pathway flux rather than as a single reaction step.[365]

Multicellular eukaryotes often control their temperature, which influences metabolic flux. Do microbes alter temperature to change flux?[141]

13.3   Genetic Drift

This section notes a common constraint on design forces. Evolutionary processes are inherently stochastic. As population size declines, stochasticity in reproduction between alternative traits may overwhelm any fitness differences between those traits.[71]

Put another way, stochastic genetic drift imposes a constraint on the potential for weak forces of design to shape traits. As always, comparative predictions give the most insight. For example, Lynch[254] showed that a reduction in population size and the associated increase in genetic drift have shaped many aspects of genomes.

Lynch considers genetic drift as a nonadaptive force. I agree. In this book, I typically label nonadaptive forces as *forces of constraint.*

For most problems discussed in this book, the forces of design are likely to overwhelm the weak constraining force of genetic drift. However, for the flux control of individual metabolic steps, drift may sometimes be important.

The challenge is to formulate comparative predictions. How do changes in population size and genetic drift alter expectations for the control of enzyme abundance levels? How do changes in robustness that protect against perturbations alter the relative strength of design forces and genetic drift (p. 107)?[125,134]

I leave those important questions and return to my main goal, clarifying design forces and the best ways to study those forces.

13.4   Challenges in Control Design

Metabolic components must adjust to each other. The overall metabolic system forms a major part of the environment for each component.

Functions of the metabolic system include sensing information, filtering out false signals, correcting errors, speeding adjustment, and enhancing stability. It can be difficult to match the particular biochemical mechanisms of metabolic components with their functional attributes at the system level.[227]

Sometimes we can think directly about component design. For example, the ATP–ADP disequilibrium provides a focal point for contrasting the efficiency of free energy capture with the use of disequilibria to drive growth. We can study how different environmental challenges alter the balance between efficiency and growth.

However, it can be difficult to match abstract system aspects of metabolic control to observable biochemical components. For example, error-correcting feedback can maintain overall system homeostasis, balancing allocations to stress resistance, maintenance, and growth. How do we relate those system-level controls and functions to the biochemical component traits that we can measure?

With those difficulties in mind, the next section lists a few problems of flux control that arose in prior sections. The last section considers the broader challenges and prospects for studying the design of control systems. The key is to focus on functional aspects, such as sensing, filtering, correcting, speed, and stability (Chapter 7).

13.5   Problems of Flux Modulation

DRIVING FORCE VERSUS RESISTANCE

The prior sections raised several problems.

- Changes in reactant concentrations alter driving force. Changes in enzymes and spatial barriers alter resistance. What conditions favor controlling flux by modulating force or modulating resistance?

- Low resistance caused by excess enzymes or other causes typically leads to near-equilibrium flux. Flux sensitivity to small changes in driving force increases as resistance drops.

- On the benefit side, near-equilibrium flux reduces the loss of free energy and raises sensitivity to force-altering changes in reactant and product concentrations, providing a fast and simple way to modulate flux.

- On the cost side, flux slows near equilibrium, product inhibition stops or reverses flux, and low resistance may require costly production of enzymes. How do differing conditions alter the weighting of these costs and benefits for near-equilibrium flux?

- Far from equilibrium, flux becomes sensitive to small changes in enzyme activity, or spatial barriers, or whatever is resisting flux. Small changes in reactant concentrations have little effect on flux. Reactions that are far from equilibrium dissipate a lot of free energy.

- When is it advantageous to regulate reactions near or far from equilibrium? How do the biophysical mechanisms that influence metabolic concentrations and resistance properties affect the design of flux control?

These problems of biochemical control form a broad area of research. Most of the work focuses on reaction dynamics and biophysical mechanisms that alter flux.[9,100,307,371,429]

CELL SIZE AND PROTEOME SIZE

Microbial cell and genome sizes vary widely.[194,206,237,325,373,391,415] Size correlates with environmental attributes and with many cellular traits, including lifespan, growth rate, and abundance. Size may influence metabolic flux.

- Size alters the ratio between membrane surface area and cell volume, which affects opportunities for membrane-based reactant gradients and for internal diffusion barriers.

- Cell volume may limit proteome size, which influences tradeoffs between the abundances of transcription factors, enzymes, and modifiers of enzyme activity.

- Fast cellular growth correlates with larger cell size. How does increased cell size alter the modulation of metabolic flux? Does a rise in proteome size and the opportunity for more proteins shift control in predictable ways?

- Larger cells for the same genome have more room for proteins. Does varying cell size reveal the relative importance of limited proteome space versus the limited genomic coding capacity?

- Advancing technology will improve measurements of kinases, transcription factors, and other proteome components. How do the forces of design shape allocations to these various classes?

- For example, do particular environmental changes favor greater response speed, relatively more kinases, or enhanced allosteric modification?

13.6   Limitations and Prospects

We can often measure how things change within cells. The problem of design concerns understanding why they change.

SCALE

The scale at which we measure often does not match the scale at which function arises. Function has to do with how cellular traits affect components of fitness. To understand design, we must match how a change in a cellular trait alters the various components of fitness.

Consider an example. We can often work out how one molecule affects another. An increase in A may enhance or repress B. Similarly, B may enhance or repress A and also affect C. The six paths between the three molecules form a little network.

If we code each path as plus, minus, or no effect, then there are 13 possible network motifs between three molecules.[9] It turns out that cells use some motifs much more often than others.

We can think of each motif as a little input-output machine. An input alters the dynamics of one molecule, such as its production or decay rate. That input-induced change shifts the molecular abundances of the network nodes, creating the output consequences of the input.

Each motif has different input-output properties. The properties include such things as how a motif amplifies an input signal, filters noisy inputs when producing outputs, or keeps the abundances of the three molecules in relative balance by feedbacks.

These facts tell us how cells deploy biochemistry to make component input-output modules. It is a bit like how computer components process electron flow to create particular logical operations. Cells use those small-scale biochemical modules to build larger networks that execute cellular control programs.

LIMITATIONS

Large-scale computer programs depend on their small-scale logical components. But we cannot understand the design of computer programs to achieve particular real-world functions by knowing only how the underlying components manipulate electrons to create logical operations.

Similarly, we cannot understand the design of cellular control to influence fitness components by knowing only how small biochemical motifs process chemical input-output operations.

Many people have understood the need to match cellular control traits to fitness components.[190] A few studies have linked cellular traits to growth rate or yield efficiency. Focus on those two fitness components arises because common methods can measure them in the laboratory.

However, one cannot understand design by those common laboratory methods of measurement. Studies must extend to the broader array of environmental challenges and fitness components that matter.

PROSPECTS

The solution is always the same. Make comparative predictions about how changing environmental challenges alter fitness components and design. Figure out how to test those predictions.

For example, how does an increase in the genetic variation between competitors alter flux control? How much does the understanding of metabolic flux and biochemistry depend on such links to natural history?

This book does not develop comparative predictions for metabolic and cellular control. That development requires synthesizing observations and extending the formulation of the key control concepts in Chapter 7 to create an applicable framework. The study of cellular control design remains an important open challenge for future work.

14 Variant Pathways

Catabolic pathways create disequilibria to drive other processes. Catabolic digestion also makes precursors to build molecules. Many catabolic functions can be achieved by alternative pathways. What forces shape variant pathways?

This chapter's biochemical descriptions provide background. That background sets the stage for linking variant pathways to fitness components, such as growth rate, biomass yield, and performance under varying conditions. Changed environments alter fitness components, leading to comparative predictions about design.

The first section reviews alternative glycolytic pathways. Those pathways differ in their required enzymes, net driving force, precursors for anabolic processes, and amounts of ATP, NADH, and NADPH produced. Those variations provide an opportunity to analyze the forces of design that favor one pathway over another.

The second section lists alternative final electron acceptors. A catabolic cascade transfers weakly held electrons in food to relatively strong electron acceptor molecules, such as oxygen or metal ions. The final electron acceptor of a cascade influences the overall thermodynamic force available to drive other processes.

The third section considers weak driving force gradients between the initial food input and the final electron acceptor. Weak gradients pose design challenges because of the low available driving force and the high potential for product inhibition.

Microbes living on weak gradients process a variety of input molecules as food, use a variety of final electron acceptors, and make diverse products that other microbes can often consume. For example, the electrons of hydrogen gas may flow to carbon dioxide, creating methane on which other microbes feed. The biochemical variety of weak gradients plays a key role in the geochemical cycles of free energy flow and in the ecological interactions of biological communities.

The fourth section describes the flow of electrons between species. Indirect flow may occur when one species excretes a catabolic product that could build up nearby, creating product inhibition and stopping catabolic flux. A second species relieves the first species' product inhibition by feeding on that species' output. Electrons flow through a free energy gradient between species, following a distributed catabolic pathway.

Direct flow occurs when one cell transfers electrons to another cell. A donor species may pass electrons to a different species that acts as the first species' final electron acceptor. Alternatively, a species may transfer electrons between its cells. In that case, the receiving cell has access to a stronger final electron acceptor than the donor cell, creating a distributed electron flux gradient between cells.

The fifth section reviews alternative pathways within a single cell that process different carbon sources for food input. Variant mechanisms to shift between food sources influence fitness components in different ways, providing a good opportunity for comparative study.

The sixth section briefly summarizes cellular shifts between complex carbohydrate food sources. The great molecular diversity and highly specific enzymatic digestion of complex carbohydrates create special design challenges to cope with the vast biochemical diversity.

The final section synthesizes puzzles of design for the variant pathways. Those puzzles form the basis for future study. Progress requires explicit comparative hypotheses and empirical tests.

14.1 Glycolytic Yield

I previously described the three common alternative glycolytic pathways on p. 180. The typical storage disequilibrium yields per glucose molecule for those pathways, from eqn 12.2, are

$$\text{EMP} \longrightarrow 2\,\text{NADH} + 2\,\text{ATP}$$
$$\text{ED} \longrightarrow 1\,\text{NADH} + 1\,\text{ATP} + 1\,\text{NADPH}$$
$$\text{HMP} \longrightarrow 2\,\text{NADPH}.$$

For EMP, the 2 ATP yield describes the common and most widely observed pattern for that pathway. Describing pathways by their typical ATP yields oversimplifies but does provide a useful starting point in the search for potentially interesting patterns and hypotheses.

For example, variant EMP pathways exist that yield more ATP. In the cellulose digesting bacteria *Clostridium thermocellum* and *C. cellulolyticum*, their EMP pathways have much lower thermodynamic driving force than the typical EMP reactions.[193,310]

Associated with low driving force and slow catabolic rate, Park et al.[310] inferred an EMP yield of 3 ATP per glucose in *C. cellulolyticum*, 1 more than the standard 2 ATP for EMP.

Digesting cellulose into glucose may happen slowly, constraining the rate at which cells can take up sugar.[193,310] When uptake rate is slow, high thermodynamic driving force and the potential for rapid flux provide no benefit. Instead, greater benefit accrues for increased free energy extraction efficiency per glucose molecule, which requires reduced net thermodynamic driving force for the pathway.

The ATP–ADP disequilibrium is smaller in *C. thermocellum* than in *E. coli*.[193] This cellulose digesting species apparently has a relatively lower rate of negative entropy uptake and a slower rate of negative entropy capture in the ATP–ADP disequilibrium.

The smaller negative entropy store in the ATP–ADP disequilibrium provides less driving force for growth and other cellular processes. The smaller ATP–ADP disequilibrium also suggests that each ATP generated requires less free energy because the free energy required to make an ATP increases with the disequilibrium, a fundamental thermodynamic fact that is frequently ignored.[294]

A smaller disequilibrium can provide a mechanism to enhance the ATP yield per glucose molecule. The same mechanism applies to any disequilibrium created as part of the catabolic yield.

Matching the slow rate of cellulose digestion to the reduced glycolytic driving force and higher ATP yield is intuitively appealing. However, several plausible tradeoffs suggest that understanding metabolic design requires more careful thought and tests (p. 223).

14.2   Final Electron Acceptors

Catabolic cascades move weakly held electrons in food to relatively strong electron acceptors. The final electron acceptor influences the total free energy between a particular food molecule and the end of the catabolic cascade.

Stronger final electron acceptors increase the total free energy change. The greater that change, the more free energy there is to be captured in the storage disequilibria to enhance yield or to be dissipated to increase flux and catabolic rate.

Weak final electron acceptors lower the total free energy change. A smaller driving force often associates with reactions that are close to equilibrium, which increases the risk that an intermediate product accumulates and stops or reverses flux.[467,468] Low driving force also makes it more challenging to capture free energy in the storage disequilibria.

I mention a few alternative electron acceptors and associated pathways.[208] In aerobic respiration, oxygen provides a strong final electron acceptor, creating a large free energy gradient from glucose to its final oxidized products. That large gradient allows capturing much free energy in the ATP–ADP disequilibrium, primarily by oxidative phosphorylation through electron transport and proton motive force.

In anaerobic respiration, final electron acceptors other than oxygen terminate electron transport chain phosphorylation. Typically, the driving force in the final catabolic steps comes from an electrochemical gradient across a membrane, associated with electron flow and proton motive force.

The final electron acceptors may, for example, be metal ions or oxidized nitrogen, which can provide strong free energy gradients.[208] However, those final acceptors attract electrons less strongly than oxygen.

Archaeal methanogens use carbon dioxide as an electron acceptor to produce methane. Bacterial and archaeal sulfidogens use sulfate or elemental sulfur as the final electron acceptor. Methanogens and sulfidogens take up a variety of electron donors as food. The total free energy gradient from electron donor to acceptor is often relatively small.

In anaerobic fermentation, typically a few post-glycolytic reactions lead to the final electron acceptor, producing lactate, acetate, ethanol, or similar molecules. The free energy gradient is often small from the food electron donor to the final electron acceptor and fermentation products.

14.3   Weak Redox Gradients

Small free energy gradients relative to the demand for metabolic flux impose strong tradeoffs. Increasing flux risks thermodynamic inhibition, in which the greater concentration of reaction products at any step along a cascade may reduce or reverse flux.

Excretion of excess reaction products may relieve thermodynamic inhibition. The overflow discards potentially usable negative entropy.

Extracting free energy yield into molecular stores of disequilibrium reduces the already limited free energy potential to drive flux, further slowing the potential flux rate.

These tradeoffs suggest how the forces of design and constraint may shape metabolic traits. I briefly mention a few theories and observations.

THEORIES

Catabolic pathways with low driving force tend to operate near equilibrium. If a bit of excess final product accumulates, the pathway may suffer thermodynamic inhibition.

The final product typically arises when the catabolic flow of electrons reduces the final electron acceptor. Three alternative designs mitigate the flux inhibition caused by the accumulating final product.

First, different microbes may catabolize the same food source to alternative electron acceptors and final products. Splitting the common food source between different pathways lowers the flux of each pathway. Lower flux reduces the concentration buildup of any particular final product, partially relieving product inhibition.[161]

Second, a single microbe may catabolize the resource through a branching pathway that ends with multiple distinct electron acceptors. A branching pathway reduces the flux into each final electron acceptor, lowering the rate at which the various final products accumulate.[467] Mixed-acid fermentation and similar architectures may be examples.[26,208]

Third, a primary microbe may catabolize the initial food source to a particular final product. A secondary microbe may then feed on the primary microbe's catabolic product, lowering the concentration of that primary product. The second microbe becomes the electron sink for the first microbe, relieving the thermodynamic inhibition of the primary catabolic pathway.[162,357]

In summary, a limited free energy gradient is particularly sensitive to concentration changes in product outputs and food inputs that alter the driving force. Limited gradients may also impose strong tradeoffs between growth rate, ATP yield, and biomass yield. Those factors influence the architecture of catabolic pathways.

OBSERVATIONS

Methanogens typically have a small catabolic free energy gradient. Archaeal methanogen clades differ broadly in catabolic pathway architecture. Those architectural differences correlate with rate, yield, and sensitivity to concentrations.[407]

The most recent clade, Methanosarcinales, uses cytochromes and electron transport across the membrane to drive ATP production. The other clades do not have cytochromes. Thauer et al.[407] summarize the broad differences between clades when the overall anaerobic catabolic reaction is

$$4\,H_2 + CO_2 \longrightarrow CH_4 + 2\,H_2O. \tag{14.1}$$

H_2 is the electron donor food source that reduces the electron acceptor CO_2 to produce methane and water.

The cytochrome group relative to the cytochrome-free group typically requires higher H_2 concentration (partial pressure), produces higher ATP and biomass yield per H_2 input, and grows more slowly. Most likely, by transferring more free energy per H_2 input into the ATP–ADP disequilibrium, this group has a lower net free energy gradient to drive the catabolic throughput and growth rate.

In natural environments with H_2, the cytochrome-free group dominates, perhaps because it requires lower H_2 partial pressure and can grow faster at a given H_2 level.[407]

Dominance by the cytochrome-free group in H_2 environments probably explains why, in natural habitats, the cytochrome group mostly grows on acetate, methanol, and methylamines. Those other electron donors may provide less free energy than H_2, consistent with the idea that the cytochrome group is more efficient and can exploit weaker negative entropy sources than the less efficient cytochrome-free group.

The glycolytic pathways discussed earlier provide additional examples of weak redox gradients and their consequences.[26,310]

14.4   Electron Flow between Cells

In oxic environments, oxygen provides the strong final electron acceptor for catabolism. If one organism overflows intermediate products, another takes up those products to complete oxidation.

Many external food molecules can be the initial electron donor. The final oxygen electron acceptor creates a large redox gradient. If the full redox gradient is separated between organisms, the sequential redox gradients are typically sufficiently strong to proceed without requiring coordination between species.

In anoxic environments, final electron acceptors are often much weaker, lowering the maximum redox gradient that can be achieved. To maintain the weak redox gradient, organisms need a steady supply of the final electron sink to accept the outflow of electrons at the end of catabolism. The need to find a final electron sink may become as great as the need to find food, the initial electron donor.

BACTERIA-METHANOGEN SYNTROPHY

A weak redox gradient is particularly susceptible to product inhibition. For example, in bacteria that pass electrons to hydrogen and release H_2, hydrogen product buildup near the cells reverses the free energy gradient, stopping catabolism.

Such bacteria require some mechanism to pull outflowing electrons away from the cells, creating a strong electron sink.[251,252,265,273]

Many bacteria in anoxic environments form syntrophic relations with archaeal methanogens.[273] The bacteria degrade various organic and other high free energy electron donors into products that retain much of the free energy, such as hydrogen, acetate, and formate (Fig. 14.1).

Nearby methanogens digest those intermediate products to methane, pulling the electrons in those intermediate products away from the bacteria and allowing the bacteria to maintain their catabolic flux. The distributed electron flux across species flows down a large free energy gradient without intermediate product bottlenecks.

For example, some syntrophic bacteria oxidize propionate to make acetate and H_2 as[214]

$$CH_3CH_2COO^- + 3\,H_2O \longleftrightarrow CH_3COO^- + HCO_3^- + H^+ + 3\,H_2.$$

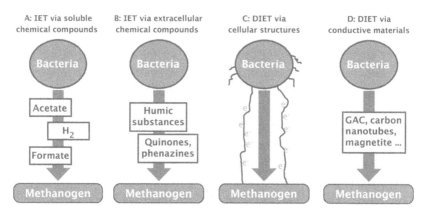

A: IET via soluble chemical compounds

B: IET via extracellular chemical compounds

C: DIET via cellular structures

D: DIET via conductive materials

Figure 14.1   Electron flow pathways from bacteria to archaeal methanogens. The methanogens use the incoming electrons plus sources of protons and carbon to make methane, as in eqn 14.1. (a) Interspecies electron transfer (IET) may happen by passing bacterial metabolic products, such as H_2. (b) Humic conducting materials in contact with both donors and recipients can transfer electrons. Extracellular organic electron shuttles, such as phenazines, may also transfer electrons between species. (c) Direct interspecies electron transfer (DIET) may happen by physical contact between cells through wire-like conducting pili or electron-transferring cytochromes in the membranes. (d) DIET via inorganic conductors. These pathways may also transfer electrons to inorganic electron acceptors, such as ferric iron. Redrawn from Fig. 1 of Martins et al.[265]

Net flux goes in the forward direction only when the concentration of H_2 remains low. Steady forward flux requires some process to remove H_2. In this case, archaeal methanogens can grow by scavenging the hydrogen to make methane,

$$3\,H_2 + \frac{3}{4}\,HCO_3^- + \frac{3}{4}\,H^+ \longrightarrow \frac{3}{4}\,CH_4 + \frac{9}{4}\,H_2O,$$

lowering H_2 concentration sufficiently to maintain forward thermodynamic driving force for the coupled system.

DIRECT ELECTRON TRANSFER BETWEEN SPECIES

A direct electron conduit between species may happen by a variety of mechanisms[251,252] (Fig. 14.1). For example, bacterial species of *Geobacter* can donate electrons via pili that act as conducting wires.

In the lab, the methanogen *Methanosaeta harundinacea* accepts electrons by direct contact with *G. metallireducens*, using those incoming electrons as their food to reduce carbon dioxide to methane.[347] In this syntrophy, electrons flow from ethanol input to methane output through a two-species catabolic pathway connected by electric pili.

Electrically conductive biofilms composed of complex multispecies communities often form on electrodes.[228,261] Apparently, many microbial species can function in long-range electron transport. In addition to electrically conductive pili, membrane-based cytochromes also seem to be important in electron flow.[184,252]

CABLE BACTERIA

Organisms in anoxic zones can significantly increase their catabolic redox gradient by connecting their electron flux to oxic environments. For example, cable bacteria in anoxic sediment donate electrons by hydrolyzing hydrogen sulfide,[151,152,210,368]

$$H_2S + 4\,H_2O \longrightarrow SO_4^{2-} + 8\,e^- + 10\,H^+. \tag{14.2}$$

The cable bacteria pass an electric current through a contiguous filament of cells that contain nickel-protein conducting wires.[45] The cellular filament terminates in an oxic zone that may be a centimeter or more away from the anoxic origin. At the oxic zone, the cable bacteria pass the incoming electron flux to oxygen to form water,

$$2\,O_2 + 8\,H^+ + 8\,e^- \longrightarrow 4\,H_2O, \tag{14.3}$$

obtaining protons by the balancing flux of ions outside the cable.[296]

Cells in the oxic zone have little anabolic activity.[151] The physiology of the cells in the oxic zone may be designed for speed of electron flux, acting solely as a wire to conduct electrons to a strong electron acceptor. The terminal transmission of electrons in the oxic zone maintains a strong electron flux and free energy gradient in the anoxic zone, where cells exploit that gradient to enhance reproduction.

Why do oxic cells provide electron flux services for anoxic cells without themselves gaining direct benefit or reproducing? I return to that puzzle below (p. 225). A later chapter discusses additional puzzles of cable bacteria design (p. 302).

14.5   Alternative Carbon Sources

A microbe exposed to multiple carbon sources typically follows a repeatable preference hierarchy of food consumption. Alternative carbon sources may fuel different growth rates, yield efficiencies, and precursor supplies for anabolic processes.

How do changing forces of constraint and design alter the preference hierarchy for alternative carbon sources? What determines the point at which microbes switch usage from one source to another? When does it pay to utilize multiple sources simultaneously? This section provides background for these questions of design.

DIAUXIC SHIFT

A microbe often uses up a preferred carbon source before shifting to use a less preferred source. A growth lag occurs during the diauxic shift between carbon sources.[380]

In the classic theory, the growth lag arises because the proteins used to digest the second source are repressed by the presence of the first source. Depleting the first source relieves the repression of proteins for the second source. It takes time to build up the second-source proteins.

I start with some details supporting this classic description. I then summarize recent studies that show two complexities in the switching between carbon sources.

First, individuals of a particular genotype may differ in their sugar usage and regulatory control. The classic diauxic shift pattern focuses on the aggregate population, ignoring the underlying individual variability. Second, genotypes may differ in regulatory control.

Variability within and between genotypes suggests that heterogeneous forces of design may shape the pattern of food usage.

Classic studies.—*E. coli* can grow on a mixture of glucose and lactose. At first, the population consumes glucose and expands exponentially. The glucose is eventually depleted. A growth lag follows. After a while, the bacteria resume exponential growth while consuming lactose.[380]

Mechanistically, feeding on glucose represses the *lac* operon and the production of the proteins required to feed on lactose. Depleting glucose relieves the *lac* operon repression. The growth lag during the diauxic shift from glucose to lactose may arise because of the time required to build up concentrations of the lactose-specific catabolic proteins.[380]

Erickson et al.[97] argued that total proteome size imposes a strong limiting constraint. In their model, the greatest growth rate arises by broad regulatory remodeling of protein expression when switching between carbon sources. Mixed simultaneous usage tends to be inefficient because of the large number of distinct proteins required for each source.

Several experiments on *E. coli* supported their model of shifting proteome allocation in response to changing availability of alternative sugar sources. Significant growth lags occurred during the proteome expression shifts.

The budding yeast *Saccharomyces cerevisiae* follows an interesting variant of the diauxic shift.[49] When provided with glucose in an oxygenated environment, the cells often glycolytically ferment the sugar to produce ethanol. Fermentation typically causes faster, lower yield growth compared with full aerobic respiration.

When most of the glucose is used up, the yeast switch to feeding on the ethanol by the slower and more efficient process of aerobic respiration. The diauxic shift from glucose to ethanol associates with the typical growth lag phase, relieves the previous glucose-induced repression of the aerobic pathways, and builds the concentration of the aerobic-associated proteins.

Yeast[425] and bacteria typically prefer glucose over sugars such as lactose, maltose, or galactose, following diauxic consumption patterns.

Variability within clones.—Those classic studies measured the aggregate consumption of populations. The implicit assumption was that all cells followed the same regulatory shifts and changes in protein expression. Recently, several studies measured the gene expression patterns within individual cells[380] in the bacterium *Lactococcus lactis* and the yeast *S. cerevisiae*.[292,389,425,437]

Lactococcus lactis follows a classic diauxic growth pattern on a mixture of glucose and cellobiose.[389] After glucose is consumed, a diauxic lag phase occurs during which there is essentially no growth, followed by consumption of cellobiose. By measuring single-cell gene expression, the authors show that only a fraction of the population switches from glucose to cellobiose consumption. The remainder does not switch and stops dividing because no glucose remains.

The higher the initial glucose concentration, the smaller the fraction of cells that switch to cellobiose after glucose is depleted. A smaller fraction of active cells on cellobiose causes a longer growth lag phase during the diauxic shift.

After the depletion of glucose and the shift to cellobiose, populations consist of some cells feeding on cellobiose and other cells in a nonfeeding and glucose-activated state. The more cellobiose-activated cells, the faster the population growth rate on cellobiose but the lower the initial growth rate if the population encounters additional glucose.

Retaining a fraction of the glucose-activated cells after glucose depletion may be a form of bet-hedging.[159,389] Because the amount of further glucose is unpredictable, fitness may be increased by a regulatory strategy that does reasonably well under a variety of future glucose encounter rates.

A lab strain of *S. cerevisiae* also expresses varying patterns of cellular heterogeneity in response to varying initial concentrations of glucose and galactose.[425] High initial glucose concentration represses the galactose pathway. Slow diauxic shift to galactose consumption follows after depletion of glucose. By contrast, low initial glucose mixed with some galactose causes all cells to induce the galactose pathway quickly.

Various intermediate initial combinations of the sugars split the population into a bimodal expression pattern. One group induces the galactose pathway before glucose is depleted, whereas the other group does not induce galactose utilization until after depletion of glucose.

With bimodal expression, the population shifts more quickly to galactose consumption after glucose depletion. The faster shift associates with a shorter growth lag. Bimodal expression may perform best when environmental sugar mixtures vary unpredictably.

Variability among genotypes.—The prior studies suggest that unpredictability favors cellular variability within clones. If so, then one might expect strains in different habitats to experience different patterns of sugar resource unpredictability and thus be tuned differently with regard to the expression of cellular variability.

Two studies of diverse *S. cerevisiae* strains support the idea of alternative regulatory tuning by genotypes.[380] New et al.[292] exposed strains to low glucose and high maltose. Some strains follow the classic diauxic shift pattern, first consuming glucose and then, after a lag, growing on maltose. Other strains follow the same glucose then maltose consumption pattern but do not have a growth lag during the shift.

Longer lag associates with faster initial growth on glucose. Shorter lag outcompetes longer lag when glucose and maltose availability alternate.

New et al.[292] suggest that short lags are a generalist strategy, gaining an advantage across different environments but, in a glucose-dominated environment, losing to the specialist strategy that maximizes growth on glucose. Isolates may vary in their genetic tendency to lag because, in nature, habitats differ in their sugar heterogeneity.

Wang et al.[437] found similar diversity in the lag between glucose and galactose usage among a wider set of *S. cerevisiae* strains. They directly measured the expression of galactose pathway genes. When consuming glucose, strains with long lag fully repress the galactose pathway. Strains with short lag induced the galactose pathway and began consuming galactose before glucose was fully consumed.

The shorter the lag after glucose consumption, the greater the expression level of galactose during glucose usage and the lower the growth rate during the glucose consumption period.

In other words, early expression of the galactose pathway is costly, reducing growth on glucose. Rapid adjustment between sugar sources reduces growth rate on the preferred sugar. Once again, more heterogeneous environments may favor a broader generalist response rather than a fast growth, specialist response.

What is fitness?—The explanations of diauxic regulatory variability assume that growth rate is the primary design attribute. For microbes such as *S. cerevisiae* grown in sugar-rich lab environments, growth rate may be the most important fitness component.

The regulatory control of feeding in other microbes may be dominated by different fitness components. Thus, it is useful to consider how other forces of design may lead to the same observed pattern of diversity.

For example, longer lags may associate with less simultaneous expression of different pathway proteins, greater efficiency, and greater total yield over the sequential consumption of a fixed amount of alternative sugars. Different environmental and demographic conditions change the relative fitness valuation of growth rate and yield (Section 17.1).

Early galactose pathway expression may induce competition for membrane space between alternative sugar transporters. The simultaneous

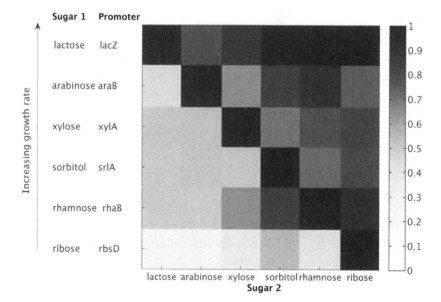

Figure 14.2 Sugar hierarchy of growth rate and gene expression in *E. coli*. Each box shows the promoter activity level associated with sugar 1 (row) when paired with sugar 2 (column). Diagonal elements correspond to single sugar treatments. From Fig. 1 of Aidelberg et al.[4]

expression of alternative transporters can provide additional targets for attack by bacteriophage and by antibacterial toxins.

Overall, growth rate typically sets the primary challenge in the lab. Nature poses a broader set of challenges.

Sugar Usage Hierarchy

Studies can infer a microbe's sugar preference hierarchy by analyzing different sugar pairs. For example, Fig. 14.2 shows that *E. coli* prefers sugars associated with higher growth rates over those associated with lower growth rates.[4,11] Individual sugars presented alone induce other pathways at low and variable levels (Fig. 14.3).

Figures 14.2 and 14.3 summarize aggregate expression levels in populations. Detailed studies show that cells vary in their expression levels,[33,212] as described in the previous subsection.

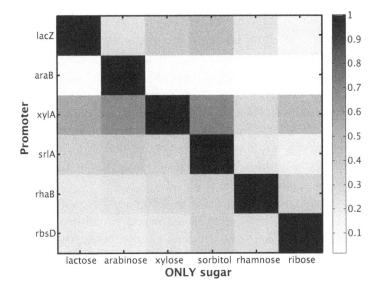

Figure 14.3 Single sugars stimulate limited expression of other sugar promoters. The column defines the sugar present. Variable shading of rows describes the expression level of other sugar promoters. From Fig. 2 of Aidelberg et al.[4]

ALTERNATIVE EXPLANATIONS AND THE NEED FOR COMPARISON

Pathways for nonpreferred or absent sugars are often partly induced. Induction happens in both pairwise and single-sugar measurements. Do those induction patterns reflect designs to increase fitness?[4]

For example, in pairwise sugar environments, it could be that simultaneous consumption maximizes growth rate. In single sugar environments, it could be that inducing pathways for absent sugars reduces growth lag if those absent sugars became available.

Alternatively, forces of design other than growth rate may dominate. Yield efficiency may be favored. Or the forces of design may vary over time and space, so that the observed traits at any point reflect complex evolutionary dynamics.

Or it could be that forces of constraint dominate. Limited proteome size may constrain the environmental sensors and transcription factors that can be used to tune expression of alternative pathways in response to environmental conditions.

Or other evolutionary forces such as mutation or drift may work against natural selection. If variant designs differ in fitness only by a small amount, then natural selection may not be able to tune design to the very best trait values.

The literature tends to emphasize consistent explanations of design rather than testable hypotheses of design. For example, the bet-hedging idea is consistent with some of the observed mixtures of sugar pathway induction patterns. But how would we know if that is the right explanation? How can we decide whether that explanation is better than many other possible explanations?

Comparative hypotheses provide the only approach to studying design. Section 17.1 develops comparative hypotheses for multiple food source usage. Before developing those comparative hypotheses, we need to have a good understanding of the observed patterns of metabolism that we wish to understand, the topic of this background chapter.

14.6   Hierarchical Usage of Complex Carbohydrates

Some microbes can break down large carbohydrates into sugars or other small molecules. Canonical metabolic pathways then process those small molecules.

Large carbohydrates are complex and diverse. Each type requires its own specific enzymes. Species vary in their hierarchical preferences for different carbohydrate forms.[137,160,332]

To cope with that molecular diversity, Bacteroidetes genomes have many polysaccharide utilization loci (PULs) to acquire and digest complex carbohydrates. Each PUL comprises multiple co-regulated genes.

Among *Bacteroides thetaiotaomicron, B. ovatus*, and *B. cellulosilyticus* WH2, each has approximately 100 PULs. The set of PULs differs significantly between species pairs.[160] *Bacteroides thetaiotaomicron* and *B. ovatus* devote approximately 18% of their genomes to PULs.[263] Other groups seem to have smaller repertoires.[374]

Why do species vary in the number of alternative pathways they encode? Why is there variation in hierarchical preferences? How do the forces of design shape the molecular regulatory mechanisms that control hierarchical feeding preferences?

The complex food molecules are often too large to be taken up directly. Instead, they must first be broken down externally. Exodigestion may be accomplished by secreting enzymes or by binding food sources at the cell surface and digesting the bound molecules. Surface digestion often associates with significant loss of the digested products before the cell can take them up.[339]

External digestion leads to public goods problems. Cells accomplishing the external digestion pay the cost for the digestion process, but the digested products are often available for neighboring cells to use. What environmental aspects favor cooperative exodigestion?

Digestion requires a sequence of steps. What situations favor different species to share in the multiple steps of distributed metabolism?

The broader forces of design inevitably modulate hierarchical preferences and exodigestion. Tradeoffs between growth rate, yield, and other fitness components must shape the biochemistry of pathways, the flux through pathways, and the regulatory controls of metabolism.

14.7   Puzzles of Design

Catabolic pathways vary because they must connect different food sources to alternative final electron acceptors. The free energy change between food inputs and electron acceptor outputs can also shape pathway design. Free energy gradients constrain flux through pathways and the amount of free energy that can be captured by the storage disequilibria.

Within those biochemical and free energy constraints, the forces of design alter pathway architecture and the tuning of flux within pathways.

Most of the literature emphasizes growth rate maximization as the primary force of design that shapes metabolic pathways.[27] Some articles also consider yield maximization, either as ATP production or biomass production per unit of food input.[317]

Rate and yield are primary fitness components. Most rate or yield explanations of design that arise from an empirical study express a consistent explanation with the data observed in that particular study.

Consistency is important. But it is also weak. Instead, we ultimately need to consider how environmental and demographic changes alter various fitness components and forces of design. And then, how do changing forces of design alter metabolic traits?

To start, it is useful to describe the puzzles of design that arise from observed patterns. A few examples follow.

PREFERENCE HIERARCHY

Microbes tend to consume available carbohydrates in a preferred order. Figure 14.2 shows *E. coli*'s preference for sugars other than glucose, the most preferred food. Higher preference corresponds to faster growth.

Human gut microbes also tend to prefer some complex carbohydrates over others. Species differ in their preference rankings.

What explains the observed preference hierarchies within species and the differences between species? Growth rate and yield are always likely candidates for strong forces of design. However, other fitness components may tune the regulatory mechanisms that control hierarchical pathway expression.

Interestingly, the way in which cells repress or jointly express alternative pathways varies within a species.[292,380,389,425,437] Temporal correlations in the availability of various foods may influence those regulatory tunings.

For example, the waiting time distribution between the consumption of a sugar and the appearance of more sugar may be important. Short wait times may favor continuously expressing the associated pathway. Long wait times may favor shutting down and then expressing again upon new stimulation. Broader temporal correlations between multiple food sources would likely have broader consequences for pathway regulation.

What fitness components most strongly influence the tuning of pathway regulation? Perhaps growth or yield. Maybe survival during periods of starvation. Or the ability to grow relative to other genotypes rather than the absolute growth rate itself.

Or, given that microbes often appear to devote much to warfare against each other,[148] perhaps attack and defense strongly shape traits. If additional free energy goes to warfare traits, then better food sources may leave growth rate or yield unchanged and instead alter success in battle.

For complex carbohydrates, initial exodigestion sometimes breaks food molecules into pieces that become available to neighbors. If so, then the value of favoring some carbohydrates over others may depend as much on the local community composition of microbes as on a cell's own internal regulation of preferences.

FLUX TUNING WITHIN PATHWAYS

Preference hierarchy concerns the regulatory tuning that controls which pathways are expressed. We may also consider the forces that tune metabolic flux within particular pathways.

For example, the EMP glycolytic pathway typically yields 2 ATP per glucose molecule. Park et al.[310] inferred a yield of 3 ATP per glucose in the cellulose digesting bacterium *Clostridium cellulolyticum*. Why does this species alter its glycolytic flux tuning to yield an additional ATP?

Cellulose breakdown may occur slowly, limiting the rate of glucose influx.[193,310] If a cell cannot increase flux and growth rate, then it may gain by reducing the net free energy driving force to match the flux limitation. Lower driving force can be achieved by devoting more of the total free energy gradient to ATP production.

At first glance, the match between slow influx, slow growth, and high yield makes sense (p. 207). And it may be so. However, several possible tradeoffs could alter tuning.

Making more glycoside hydrolase could potentially increase cellulose breakdown and glucose flow rate. Making more surface transporters could potentially increase glucose uptake rate. Greater uptake rate could potentially favor greater glycolytic driving force, increased growth rate, and reduced yield. The association between cellulose digestion and ATP yield may also depend on environmental and demographic factors.

In other words, one achieves only a limited approach to understanding design by intuitively matching difficult digestion to slow growth and high yield. We need comparative predictions and tests.

Other puzzles of pathway tuning arose in earlier sections. In overflow metabolism, what determines the balance between overflow excretion of fermentation products and full processing through oxidative phosphorylation (Section 12.2)? In futile biochemical cycles, what forces tune the dissipation of disequilibria and the generation of excess heat (p. 167)? In oxidative phosphorylation, what tunes the balance between ATP-generating efficiency and flux rate (p. 178)?

PATHWAY ARCHITECTURE

For a given food input, differences in the final electron acceptor and total free energy gradient may associate with broad pathway differences. An obvious distinction occurs between anaerobic and aerobic pathways. The much larger free energy gradient when using oxygen as the final electron acceptor associates with many additional biochemical steps in the TCA cycle and oxidative phosphorylation.

Small total free energy gradients may diversify pathway architectures. With a limited total gradient, slight variations in flux can strongly alter the efficacy of a particular architecture. That amplification of consequence may induce architecture variety to deal with various challenges.

For example, when the final electron acceptors follow soon after basic glycolytic processing, various terminal pathways and branching architectures occur. What forces cause termination in pyruvate, ethanol, lactate, or other products? Why do some terminal pathways branch to produce more than one final product?

Thermodynamic aspects may sometimes be important. For example, it may matter how pathway variants relieve product inhibition, modulate the total free energy gradient, change the free energy differences in particular biochemical steps, and alter the free energy capture in storage disequilibria.

Making comparative predictions may be challenging at the broad scale of architectural differences. Nonetheless, we need comparative predictions to go beyond intuitive matching between observed pattern and hypothesized process.

An interesting contrast in pathway architecture occurs in the archaeal methanogens (p. 210). The Methanosarcinales clade uses cytochromes and electron transport across the membrane to drive ATP production. The other clades do not have cytochromes.

The cytochrome group relative to the cytochrome-free group typically grows more slowly and has higher yield per H_2 molecule input. However, the cytochrome-free group dominates in H_2 environments, perhaps because it can grow faster at a given H_2 concentration.[407]

The cytochrome group is found mostly in habitats where it can grow on acetate, methanol, and methylamines. Those alternative food sources may provide less free energy than H_2. Perhaps the cytochrome group can exploit weaker negative entropy sources more efficiently than the less efficient cytochrome-free group. Stronger comparative predictions and tests would be useful.

Cooperative catabolism

Three examples pose interesting puzzles.

First, cable bacteria link cells to connect anoxic-zone hydrogen sulfide electron donors to oxic-zone electron acceptors. The electron flow gradient causes rapid removal of electrons from anoxic-zone cells, enhancing the free energy gradient for those cells. The oxic-zone cells pass the incoming electrons on to the final oxygen acceptor (p. 213).

The anoxic-zone cells grow and divide. The oxic-zone cells do not. If oxic-zone cells do not gain any growth advantage and instead are effectively sterile altruists, why do they express the cooperative traits that connect anoxic cells to the oxic zone?[151]

Most likely, the cells are genetically similar, favoring cooperation by kin selection. Additionally, the cells may be randomly located in anoxic versus oxic zones, which in certain conditions may favor cooperative and nonreproductive expression in those cells that happen to end up in the oxic zone.

If the cells move frequently, they may alternate between reproductive and nonreproductive phases.[151] Section 17.2 develops potential explanations into comparative predictions.

In the second example, extracellular electron shuttles accept electrons from some cells and donate electrons to other cells or to abiotic electron sinks.[250,409] Such shuttles have mostly been studied in experimentally manipulated systems. Shuttles may be quinones, phenazines, or a variety of other molecules.

In natural systems, extracellular shuttles would be available to any nearby cells. It is not clear how often such shuttles are produced and released by cells specifically for electron transport function. If it happens, then cells that make shuttles would pay the cost of production and share the benefits with neighbors (p. 311). Such publicly available resources raise interesting aspects of conflict and cooperation in design.

Third, catabolic outputs excreted by one species may be taken up as catabolic inputs by another species. For example, bacteria in anoxic environments may release acetate, formate, H_2, or other catabolic products with significant negative entropy. Archaeal methanogens may catabolize those molecules to methane and water (p. 211).

Methanogen uptake of the bacterial catabolic products relieves the bacteria of potential product inhibition. Puzzles concern how the flux of bacterial catabolic products depends on the environment, the free energy changes in various biochemical steps, and the resistance that may oppose flux at any point in the two-species pathway flow.

Pathway characteristics may be tuned to variation in environmental and demographic attributes. In addition, traits of one species may affect the population dynamics of the other species,[113] which in turn alters the flux and free energy gradient of the cooperative catabolic pathway.

15 Tradeoffs

> In order to spend on one side, nature is forced to economise on the other side.
>
> —Goethe, quoted by Darwin[74]

Tradeoffs set the basis for design. To grow faster, it would seem that resources must be taken away from some other function. If there were no cost, why would the organism not already have achieved faster growth?

But focusing solely on tradeoffs will fail to reveal design. Suppose, for example, that growth rate, yield, and defense trade off against each other. Changed conditions may reduce allocation to defense, allowing both growth rate and yield to rise.[300] If we do not measure defense, the rise in rate and yield would seem to contradict the rate-yield tradeoff.

There will always be another tradeoff not measured. One cannot win solely by collecting more data. There has to be a proper method behind the effort (Chapter 3).

Another problem is that organisms are not perfectly adapted. A cell might evolve a more efficient catabolic enzyme, capturing more free energy without costly reduction in another function. Some improvements can happen without tradeoff.

If the study of tradeoffs often fails to reveal design, then why focus on them? Because tradeoffs do play the central role in design. The problem is not the importance of tradeoffs in shaping design. The problem is our ability to infer the causes of design.

I have argued that comparative hypotheses provide the only approach to inferring the causes of design. Tradeoffs are the building blocks of comparative hypotheses. Before turning to comparative hypotheses in the following chapters, it is useful to consider in this chapter the potential tradeoffs that may be important in different situations.

We will never guess all of the important tradeoffs when developing theory. And we will never measure all of them in empirical study. But the more tradeoffs we know about and the more of those tradeoffs that we use to develop comparative predictions, the closer we will come to understanding the forces that shape organismal design.[20,104,209,379]

Typically, one assumes that a tradeoff exists and then considers what predicted consequences follow from that assumption. How does one test for the tradeoff itself? As always, attempting direct measurement provides a limited signal about the forces of design.

Testing a comparative hypothesis provides the way forward. How do changing environmental conditions alter the likelihood that a particular tradeoff influences design? To develop such comparative hypotheses, one must have a good sense of the possible tradeoffs.

This chapter lists some tradeoffs for metabolic traits. These examples encourage thought about what might be important. The examples also provide the basis for developing broader lists, which would help to locate the boundaries for possible explanations of design. The actual is always a subset of the possible.

15.1   Biophysical Constraints and Cellular Allocation

How do physical limits, such as membrane surface area, impose tradeoffs between alternative functions? How do cells split investment between the structures needed for growth, such as ribosomes, and productive investment in actual growth, such as making new proteins?

LIMITED SPACE

Cells allocate their limited membrane surface area and their limited cellular volume to alternative functions. Those biophysical limits impose particular tradeoffs.

- Limited membrane surface area imposes tradeoffs. For example, food typically passes through membrane transporters. Oxidative phosphorylation uses the membrane to conduct electron transport. At high throughput, membrane space imposes a tradeoff between food uptake and electron transport.[399]

- Limited cellular volume constrains the total number of proteins, imposing tradeoffs. For example, the abundance of transcription factors that regulate cellular function trades off against the abundance of structural proteins that carry out cellular function.[329]

- Cell crowding slows diffusion and limits biochemical reactions.[16,359] To overcome slow diffusion, reaction cascades may colocalize into phase-separated compartments.[37] Tradeoffs likely arise in the maintenance of spatial partitioning within limited cellular volume. The implications for cellular design remain an open problem.[5,25,238]

RESOURCE ACQUISITION

Cells must acquire various nutrients. Tradeoffs occur between finding, uptake, and efficiency.[247] Tradeoffs also arise between the uptake of different nutrients.

- Exploration for new food resources by motility trades off against efficient exploitation of local food sources.[164,395,402] Tradeoff strength increases as the cost rises for motility and the benefit rises from the patchiness of food distribution.

- A related tradeoff occurs between costly motility for individual cells and the beneficial tendency to be at the edge of a growing colony. Motility in colonies segregates cells by their speed, potentially colocalizing cells with synergistic traits.[474]

- Patchy environments favor motility because of the high gains for finding new food patches. However, in resource-poor environments, rich patches tend to be rare, favoring the cost savings associated with relatively low motility.[299]

- Nutrient transporter affinity trades off against cost.[103] In rich environments, low affinity and low cost transporters maximize nutrient uptake rate per unit cost. In poor environments, higher affinity and higher cost transporters achieve greater nutrient uptake efficiency, as in the widely conserved ABC system.[299]

- Catabolic pathways for different food sources associate with matching uptake systems.[81] Tradeoffs may occur between alternative uptake systems.

ALLOCATION TO ALTERNATIVE ASPECTS OF GROWTH

Tradeoffs occur between growth-related functions.[156,240,440,464]

- The rate of protein production trades off against the yield efficiency of resource usage.[343] Species with more ribosomal RNA operons make proteins faster, grow faster, and have lower biomass yield per unit carbon uptake.

- Increasing resources favor protein production rate over yield. In 1167 bacterial species, ribosomal RNA copy number increases with traits that are common in resource-rich environments, such as chemotaxis and larger genome size.[343]

- Allocation to alternative proteins varies with environmental challenge, suggesting broad tradeoffs.[188,240] Changes in the marginal costs and benefits of flux through different pathways may explain cellular shifts in proteome allocation to different cellular functions.[188]

- In yeast, mitochondria mediate a growth rate versus yield tradeoff. The mitochondria change from primarily a biosynthetic anabolic hub during rapid fermentive growth to primarily a catabolic hub during slower and more efficient respiration.[83]

- Limitation of essential elements induces tradeoffs between pathways. For example, in *E. coli*, phosphate limitation favored a mutant with lower TCA cycle flux. That mutant had reduced fitness under carbon limitation.[32]

- Rapid growth requires more phosphorus for ribosomes, whereas other life histories demand relatively higher availability of carbon or nitrogen.[95,147,211,245,281]

FREE ENERGY STORES FOR DELAYED FUNCTIONS

Molecular storage includes glycogen, trehalose, polyphosphates, polyhydroxyalkanoates, various lipids, and occasionally polypeptides.[208,378]

When lack of essential nutrients limits current growth, microbes may gain by storing available nutrients and free energy for future benefits. Alternatively, tradeoffs between current and future benefits may favor storage. Possible future benefits include the following.

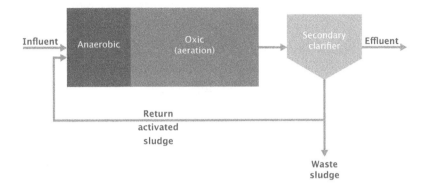

Figure 15.1 Biological phosphorus removal process. Redrawn from Fig. 17 of Curtin et al.[72]

- Many yeast and bacteria store glycogen to use during starvation conditions.[450]

- Storage may allow rapid transition between regulatory states. In *E. coli*, rapidly accessible glycogen stores reduced the lag time during shifts between alternative carbon sources.[372]

- Alternative glycogen molecular structures may trade off stability versus rapid accessibility.[438] Stability favors starvation resistance. Accessibility favors rapid regulatory changes. Arguments continue about how glycogen's structural branching influences long-term molecular stability versus short-term accessibility of free energy.[438,450]

ALTERNATING ENVIRONMENTS FAVOR CYCLES OF STORAGE AND GROWTH

The microbiology of human-engineered water purification treatment provides a good example of such cycles.[72,88,297]

Figure 15.1 shows a common process. Waste often contains common products of glycolytic fermentation, such as acetate. The first purification step begins with anaerobic bacterial activity. In the absence of oxygen or other strong electron acceptors, the bacteria cannot use acetate to generate a free energy gradient for growth.

Instead, cells draw on internal stores to generate usable free energy. They may derive glucose from stored glycogen or cleave phosphate

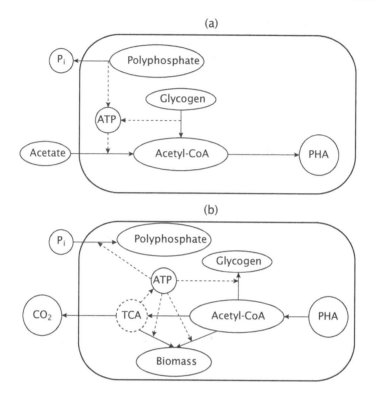

Figure 15.2 Metabolic pathways of storage and growth for common phosphate-accumulating bacteria in waste processing systems. (a) During the initial anaerobic phase, bacteria take up volatile fatty acids, such as acetate. Free energy from polyphosphate or glycogen stores drives transformation of the fatty acids into PHA storage. (b) During the subsequent aerobic phase, free energy obtained by oxidizing the stored PHA drives biomass production and the building of polyphosphate or glycogen stores. Metabolic variations occur among different species. For example, glycogen-accumulating bacteria use relatively little polyphosphate and so do not remove phosphates from the waste. When carbohydrates are present in the initial waste, anaerobic fermentation often transforms those carbohydrates into volatile fatty acids. From Fig. 2 of Dorofeev et al.[88]

bonds in stored polyphosphate. The free energy from those reactions drives uptake and transformation of acetate into polyhydroxyalkanoate (PHA) stores (Figs. 15.2a and 15.3b).

After an anaerobic period, the treatment process aerates the waste. Cells derive usable free energy by oxidizing the PHA stored during the anaerobic phase. The free energy from PHA drives growth and biomass

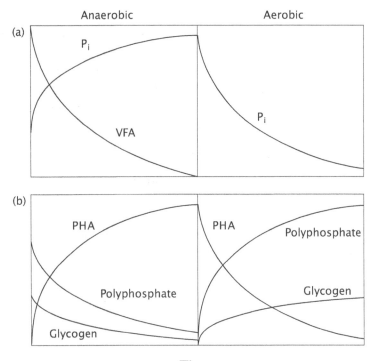

Figure 15.3   Bacterial uptake and storage of key nutrients over the cycle of anaerobic and aerobic environments in phosphate-removal processes. (a) Extracellular concentration levels of phosphate and volatile fatty acids (VFA), such as acetate. (b) Internal cellular stores for phosphate-accumulating species. In glycogen-accumulating species, the external phosphate changes relatively little and the internal polyphosphate remains low. Instead, glycogen storage falls more rapidly during the anaerobic phase and builds more rapidly during the aerobic phase. From Fig. 1 of Dorofeev et al.[88]

production. Cells also use the free energy from PHA to replenish their stores of polyphosphate or glycogen (Figs. 15.2b and 15.3b).

The bacterial biomass can be removed as waste sludge after one or more anaerobic-aerobic cycles. Sludge removal discards any stored carbon or phosphate, purifying the water.

One practical challenge concerns competition between bacteria that build polyphosphate stores and bacteria that build glycogen stores. The phosphate-accumulating bacteria provide efficient removal of phosphorus from waste. The glycogen-accumulating bacteria leave much of the phosphorus in the waste (Fig. 15.3). Tuning of the purification process requires finding the steps and environmental setpoints that favor phosphate-accumulating over glycogen-accumulating species.[351]

This system provides an excellent model for the study of metabolic design.[383] Although artificial, it captures the essence of alternating environments, the benefits of storage, the thermodynamic constraints imposed by varying electron acceptors, and the shaping of long-term fitness over a full demographic cycle.

Several tradeoffs likely influence the fitness of particular genotypes and the competition between genotypes and species.[383]

- During the aerobic phase, cells trade off biomass production against building polyphosphate or glycogen stores.

- Polyphosphate and glycogen production draw on the same ATP supply. A tradeoff may occur between building those alternative storage molecules.

- For example, polyphosphate and glycogen provide different precursors for other metabolic processes. They make different demands for uptake and membrane space. They likely have different redox characteristics, temperature sensitivities, pH sensitivities, and other chemical properties.

- During the anaerobic phase, uptake of volatile fatty acids may trade off against uptake of food sources that could drive glycolytic fermentation and growth. Limited membrane space or limited free energy to drive active transport may impose the tradeoff.

Changes in Limiting Resources Cause Changes in Tradeoffs

Tradeoffs between alternative allocations often depend on context. Accounting for context improves comparative tests.

- When lack of particular nutrients limits growth, cells may store available nutrients for future benefits.[88,450] Building storage molecules may trade off against current maintenance, repair, and survival.

- By contrast, when available nutrients allow growth, the future bene-fits of storage may trade off against current growth.

- Restricted phosphorus limits ribosome and protein production.[95] Limited protein imposes a strong tradeoff between making proteins for new growth and making proteins for other functions, such as cellular regulation. Abundant phosphorus may relieve proteome limitation and reduce the importance of protein allocation tradeoffs.

- Tradeoffs that may dominate in one context may be weak or absent in another context. For example, limited free energy from available food sources means that each free energy consuming function trades off against all other free energy consuming functions.

- By contrast, free energy may be available in excess compared to an elemental limitation, such as nitrogen. Elemental limitation changes the particular tradeoffs that dominate design.

- In general, different elemental and resource limitations impose dif-ferent costs and benefits for various cellular functions.[66,67,101,104,249] Changes in costs and benefits alter the relative strength of tradeoffs between different functions.

15.2   Exploration versus Exploitation versus Regulation

Cells explore their environment to gain information and exploit their environment to gain resources. Cells use information obtained from exploration to regulate exploitation. Tradeoffs arise between exploration, exploitation, and regulation.

CELLULAR CONSTRAINTS

Genome size, cell size, and limited resources impose tradeoffs.

- More resources devoted to exploratory information-gathering im-ply fewer resources devoted to exploitation. For example, sensor abundance may trade off against enzyme abundance.

- More resources devoted to regulation imply fewer resources devoted to information and exploitation. For example, transcription factors may trade off against information and exploitation proteins.[329]

- Genome and cell size may trade off against exploration, exploitation, or regulation. In oceanic surface prokaryotes, small genomes correlate with nutrient-poor environments, slower growth, and fewer genes associated with adjusting to alternative conditions.[463]

- Small genomes may benefit in poor environments by more efficient maintenance and growth. Those benefits may trade off against a reduced proteome for information and regulation.

EXPLORATION VERSUS EXPLOITATION

Cells explore by moving through the environment and by using sensors of internal and external environmental states. Costly exploration reduces exploitation efficiency.

- Exploration by motility trades off against exploitation and growth efficiency (p. 229).[164,299,395,402]

- In yeast, exploration by glucose sensors trades off against the exploitation efficiency of growth for already deployed glucose uptake receptors and catabolic pathways.[465] Yeast have a broad range of sensors that regulate metabolism.[67]

EXPLOITATION VERSUS REGULATION

Cells adjust by altering regulatory controls. Faster and more extensive regulation may reduce exploitation efficiency and growth rate.

- Fast mRNA decay speeds transitions to new cellular states by increasing message turnover. Fast mRNA decay also burns resources and degrades exploitation. Excess resources reduce the exploitation cost, favoring more rapid mRNA decay and regulatory adjustment.[285]

- Growth rate trades off against the rate of regulatory change when switching from one food source to another. Different regulatory mechanisms mediate this tradeoff (Section 14.5).[29,380]

- Greater mRNA transcription rate imposes costs, degrading exploitation. Greater transcription rate also reduces stochastic fluctuations in protein production, improving regulatory precision. Observations across different genes and species follow this tradeoff between exploitation cost and regulatory precision.[174]

DUAL FUNCTION

A membrane uptake receptor may also act as an information sensor.[67] In general, dual function reduces tradeoffs if a molecule can be tuned simultaneously for alternative actions.

Dual function does not contradict the central role of tradeoffs. Instead, the forces of design favor simultaneous improvement over tradeoffs, until further dual improvement cannot be achieved. At that point, improvement in one dimension often demands loss in another dimension. The tradeoff has returned (Fig. 3.1).

15.3 Thermodynamics and Biochemical Flux

Thermodynamic driving force and resistance to chemical transformation determine flux. The mechanisms that modulate driving force and resistance impose tradeoffs between components of success.

REGULATORY CONTROL TRADEOFFS

Various cellular mechanisms influence flux.[100] Gain from modulating flux may trade off against loss in other components of success.

- Some glycolytic reactions have excess enzyme capacity and run near equilibrium.[310] Excess enzyme allows rapid increase in flux to meet the demand from increased resource flow. The benefit of flux adjustment trades off against the cost of excess enzyme.

- Raising enzyme concentration for a particular reaction and lowering concentration for another reaction may be a slow process. The efficiency gained by adjusting enzyme concentrations trades off against the slow response to changed conditions.

- Cells modify small parts of enzymes. Enzyme modification changes the resistance of reactions. Covalently modified enzymes provide cheap and fast flux control but may be less specific and efficient than modulating the concentrations of custom-designed enzymes.

- Thermodynamic driving force rises with greater reactant concentrations and smaller product concentrations. Stronger driving force increases flux but also dissipates free energy, lowering efficiency.

ADDITIONAL MECHANISMS THAT ALTER FORCE OR RESISTANCE

This subsection lists a few examples.

- Removing products to increase driving force may require costly processing or excretion of otherwise useful molecules.

- Separation of reactants modulates resistance. Biophysical changes to benefit one reaction may impose costly alteration of resistance for other processes.

- For example, lowering resistance by increasing membrane flux for a reactant may also increase the flux of toxins.

- Membranes impose resistance, maintaining chemical gradients and potential driving force.[35,86] Membranes also provide structural, sensor, and active transport functions. Altering a particular membrane property may influence the costs and benefits of several functions.

RATE VERSUS YIELD

Two commonly discussed tradeoffs arise.

- First, enhanced thermodynamic driving force increases the rate of a reaction. That increased rate associates with a reduced free energy yield that can do useful work.[47,317,444] This rate versus yield tradeoff arises from fundamental aspects of thermodynamics.

- Second, cellular growth rate often trades off against biomass yield. This rate versus yield tradeoff arises in observations.[257,326]

- The two tradeoffs may be linked by assuming that cellular growth rate depends on biochemical reaction rates and that biomass yield depends on the usable free energy yield of reactions. Those assumptions often makes sense. But they are not guaranteed.

RATE VERSUS THERMODYNAMIC INHIBITION

When reaction products build up in concentration, thermodynamic driving force declines. In other words, reaction products thermodynamically inhibit flux.

- The benefits of increasing reaction rate may trade off against the costs of relieving product inhibition to maintain rapid flux. This tradeoff shapes overflow metabolism (Section 12.2).

- When driving force is low, small increases in product concentrations greatly reduce flux. Thus, low driving force's gain in free energy efficiency trades off against greater sensitivity to product inhibition.

- Fast reactions incur costs to maintain high flux, whereas slow reactions incur costs to maintain positive flux.

- The total driving force over a metabolic cascade may be nearly constant. Greater driving force in one step trades off against lower driving force in another step. Reduced driving force of a step raises the risk that product inhibition blocks the entire cascade.

- In a cascade, a stronger final electron acceptor increases the total driving force. More steps reduce the force per step. Driving force and product inhibition impose tradeoffs when altering the final electron acceptor or the architecture of the cascade.

RATE VERSUS REDOX IMBALANCE

Cells maintain ratios of reducing and oxidizing compounds. Perturbed redox ratios interfere with regulation of biochemical processes.

- Rapid catabolic flux trades off against redox imbalance. In $E.\ coli$[423] and yeast,[186,424] rapid flux creates an excess NADH–NAD$^+$ redox imbalance. Lack of NAD$^+$ limits the necessary electron acceptor to maintain flux through the TCA cycle.

- Electron transport takes electrons from NADH, yielding NAD$^+$. Rapid upstream catabolic flux may produce NADH faster than electron transport can use it, creating an NADH–NAD$^+$ imbalance (p. 167).

- Futile cycles dissipate the free energy in ATP–ADP and NADH–NAD$^+$ disequilibria.[348] Futile cycles suggest a tradeoff between some beneficial process that creates redox imbalance and the potentially costly dissipation of free energy to restore redox balance.

STRONG ELECTRON ACCEPTOR VERSUS OXIDATIVE DAMAGE

Oxygen is a strong electron acceptor that provides powerful thermodynamic driving force to catabolic cascades. Reactive oxygen species (ROS) also cause uncontrolled oxidative damage. Cells use various mechanisms to mitigate ROS damage.

- Strong electron acceptors trade off enhanced driving force against increased oxidative damage.[467]

- Growth rate trades off against ROS mitigation efficiency.[12]

15.4   Fitness Components and Life History

Life history analysis separates long-term success into fitness components. Components include reproduction, survival, and dispersal.[393] In microbes, we typically equate *reproduction* and *growth*. By contrast, multicellular *growth* usually describes size increase from birth.

Life history emphasizes tradeoffs. Faster reproduction reduces survival. More dispersal lowers reproductive rate.

Fitness components may be divided into subcomponents by habitat, age, or other factors. Survival in food-limited habitats may trade off against reproduction in food-rich habitats. Dispersal out of old resource patches may trade off against reproduction in young resource patches.

Age-specific tradeoffs are important in life history. For example, reproduction early in life often trades off against reproduction late in life. In microbes, a lineage's reproductive age is relatively younger at present and relatively older at a later time.

Lineages vary reproductively over time by accumulating damage or by intrinsic fluctuations, causing variation in age-specific fitness.[330] Senescence occurs when age-specific fitness declines with age.[3,345]

The following examples highlight tradeoffs between current reproductive rate and various fitness components. Typically, current reproductive rate associates with a lineage's metabolic traits and growth rate.

GROWTH RATE VERSUS BIOMASS YIELD

A faster rate of cell division trades off against the biomass yield per unit of food intake. Prior sections described many examples of this tradeoff and possible underlying mechanisms.[65,97,103,104,207,246,257,326,343,369,463]

GROWTH RATE VERSUS SURVIVAL

Greater current reproduction may trade off against lower survival. Reduced survival decreases future reproduction.

- Faster growing *Schizosaccharomyces pombe* yeast cells die at a faster rate.[291]

- Among 16 bacteriophage species that attack *E. coli*, thinner capsid surfaces and more tightly packed genomes correlated with faster growth and lower survival.[79]

GROWTH VERSUS MAINTENANCE AND STRESS RESISTANCE

Faster growth may trade off against maintenance and stress resistance. Such mechanistic links may explain the tradeoff between growth and survival.

- In *E. coli*, a few transcription factors act as master regulators of growth and maintenance. Stressed cells reduce growth and upregulate maintenance. Competition between these master regulators for access to RNA polymerase and transcription may mediate growth versus maintenance tradeoffs.[266,301]

- RpoS is a master transcriptional regulator of *E. coli*'s stress response. Greater stress resistance associates with reduced catabolic processing of diverse carbon sources. Widespread polymorphism in the *rpoS* gene or its expression level occurs among strains, apparently tuning stress resistance versus growth to local conditions.[102]

- Variable membrane permeability trades off nutrient uptake against susceptibility to oxidative stress.[102,177]

- In *S. cerevisiae*, slower growth and higher biomass yield correlate with better stress resistance. That correlation occurs for different environments, nutrients, or mutations. Genomic knockout analysis reveals many mutations that mediate the tradeoff between growth and stress resistance.[466]

- Similarly, in *Schizosaccharomyces pombe*, a high-flux variant of the glycolytic enzyme Pyk1 associates with fermentation, fast growth, and decreased oxidative stress resistance. The low-flux variant causes a catabolic switch to respiration, slower growth, and greater stress resistance.[202]

GROWTH RATE VERSUS AGING RATE

Senescence occurs when survival or other fitness components decline with age. Lineage age can be measured as a particular amount of time or a particular number of cell divisions. The aging rate is the rate of decline in age-specific fitness.

- Aging in a lineage associates with accumulating cellular damage. In some species, damaged molecules segregate asymmetrically during cell division. One cell suffers a decline in age-specific fitness. The other cell maintains fitness or is rejuvenated.[2,225,241,330]

- Greater stress-induced damage enhances asymmetry's growth benefits.[422] If greater asymmetry imposes higher costs on other fitness components, then the increasing growth benefits of greater asymmetry trade off against those higher costs.

- In *E. coli*, stress resistance mediates a tradeoff between aging and growth. For example, increased expression of the general stress pathway reduces the aging rate and lowers the growth rate.[461]

TRADEOFFS WITH DISPERSAL AND DORMANCY

Most resource patches eventually disappear. New ones open up. A lineage's long-term success depends on colonizing new locations.

In a stable habitat, a lineage gains by dispersing a fraction of descendants to compete with others, reducing competition with itself.[111,169]

Dispersal may occur over time rather than across space. For example, dormant spores reduce activity and protect themselves to survive the journey to a later time.[229]

The benefits of colonizing different locations or later times trade off against various costs.

- Dispersers typically pass through time or space nonreproductively.

- Dispersal often exposes microbes to mortality risk.

- Active movement requires resources.[344] Dormant quiescence requires storage and maintenance. Those dispersal-related resources lower allocation to other functions.[229]

- Dispersal trades current reproduction for future reproduction in a different location. The reproductive value of growth in different patches depends on demographic processes. Reproductive value translates a unit of reproduction in different classes into a common valuation for long-term contribution to the population.[61,122,405]

- Similarly, dormancy trades current reproductive opportunity for later opportunity. The reproductive value of current versus later reproduction depends on demography.

- Decaying conditions favor allocation to distant opportunities.

15.5 Warfare versus Productive Traits

Microbes attack competitors and defend against assault. Attack and defense trade off against productive growth, survival, and dispersal.

- Greater cell membrane permeability may enhance resource uptake and growth. That growth benefit trades off against greater susceptibility to antibiotics and other attack molecules that pass through the membrane.[177,318]

- Many resource uptake receptors provide the site of attack by bacteriophage. Increased resistance to attack trades off against reduced resource uptake.[104]

- A significant fraction of microbial genomes may be devoted to attack and defense. Limitations on total coding, transcription, and proteome size impose tradeoffs between warfare and productive functions.[146,148,157]

15.6 Cooperative Traits

Analyses of metabolic design typically focus on the success of a clonal lineage. How fast does the lineage grow? How efficiently does the lineage transform resources into biomass?

However, many studies identify microbial traits that reduce a lineage's success while enhancing the success of other lineages.[441] The tradeoff between growth rate and biomass yield provides a good example.[317]

A lineage that reduces its growth rate leaves more resources for neighboring lineages, including itself. We must weigh a lineage's cost for reduced growth against the yield benefit provided to neighboring lineages.

Ultimately, natural selection can favor traits that directly reduce the success of a lineage and indirectly increase the success of similar traits in other lineages. Hamilton's[166] classic tradeoff compares a lineage's direct cost, c, to the benefit for other lineages, b, weighted by the genetic similarity between lineages, r.

When a lineage is by itself, then all neighbors are the lineage itself, and $r = 1$. As neighbors become increasingly different from the focal lineage, r declines.

A cooperative trait is favored when the similarity-weighted benefit outweighs the direct cost, $rb > c$. For example, when considering slower growth, cooperation depends on how the similarity-weighted yield benefit trades off against the direct growth cost.

An alternative aspect of similarity can also be important.[122,405] Instead of measuring how a lineage's cooperative trait benefits genetically similar neighbors, we can measure how a lineage's cooperative neighbors provide benefit to the focal lineage (Section 5.2).

In this second case, a lineage that enhances group success has similarly cooperative neighbors in proportion to r. We once again get the condition $rb > c$ for a cooperative trait to be favored. The distinction is that, in the first case, benefits flow from our focal cooperative lineage to genetically similar neighbors, whereas in the second case, benefits flow from phenotypically similar neighbors to our focal cooperative lineage.

The distinction is important because, in the second case, the cause of similarity between neighbors does not have to be genetic. For example, a lineage could have cooperative neighbors of a different species.[113] Thus, similarity does not necessarily depend on kinship or pedigree, although common ancestry can be one cause of similarity.

Cooperative traits may also gain by payback through an ecological loop.[73] For example, a lineage may excrete a useful metabolic product, the loss of which directly reduces the lineage's success.

That costly excretion may ultimately be favored by a synergistic feedback loop. Suppose an excreted product enhances the growth of another lineage that, in turn, excretes a product beneficial to the first lineage. In that case, the initial secreting lineage trades off an immediate cost for a later return benefit.

Cooperative tradeoffs have been widely discussed in the microbial literature, with many examples. The point here is that we must consider such tradeoffs when trying to understand the design of metabolic traits. In essence, sociality shapes biochemical and metabolic design.

Similarity Selection

A lineage may trade the lost success from expressing cooperative traits for the benefits received from similar neighbors.

- Reduced growth rate imposes the cost of slower reproduction. Neighbors benefit from the resources left unused. In a yeast experiment, the relatively high similarity-weighted benefit from neighbors, rb, outweighed the cost of reduced growth, c, favoring a low growth and high yield strain.[258]

- Oil emulsions limit diffusion and keep beneficial unused resources close to a slow-growing cell. The smaller neighborhood also increases the similarity, r, between nearby strains. In laboratory evolution experiments, greater r favored slower growth and higher yield,[21] with additional tradeoffs possibly playing a role.[401]

- Microbes often secrete exoenzymes to break external food molecules into smaller pieces.[335,473] The secreting cell pays the cost of enzyme production. All nearby cells gain the benefit of extracellular digestion. A producing cell trades off the similarity-weighted digestion benefit, rb, against its production cost, c.

- In marine bacteria, extracellular enzymes break chitin into smaller pieces. Cooperative secretion of those enzymes decreases when diffusion of usable pieces exceeds uptake by nearby cells.[92] Fast diffusion increases the spatial scale of benefits and reduces r, the similarity of neighbors that share benefits.

Cooperative Cross Feeding

A lineage may trade its direct immediate success for a later return benefit through an ecological feedback loop.

- In an experiment, *Salmonella enterica* evolved costly secretion of methionine to enhance the growth of a methionine-dependent *E. coli* strain. The enhanced growth of *E. coli* increased its secretion of waste acetate, on which *S. enterica* depended for its growth.[170]

- A follow-up experiment supplied lactose, which catabolically splits into glucose and galactose. *Escherichia coli* evolved to secrete galactose, boosting the growth of *S. enterica* and thus increasing the methionine supply on which that *E. coli* strain depends. This second step established mutually cooperative cross feeding.[171]

15.7   Timescale Tradeoffs

A beneficial trait on one evolutionary timescale may be a costly trait on another evolutionary timescale. Timescale tradeoffs pose one of the great challenges in the study of design.[133]

Consider a canonical problem of microbial life. A patch opens up with a fixed amount of resources. A colonizing microbe lands on the patch and begins to grow. A mutant arises that grows faster than its progenitor lineage but is less efficient at converting resources into biomass.

The mutant overgrows its competitors, dominating the population. Growth depletes local resources, ending the patch life cycle. Over the life cycle, some cells disperse and colonize new patches.

Over the short timescale of a cellular generation within the patch, direct competition favors rapid growth.[236] Over the long timescale of the full life cycle, dispersal typically favors greater biomass production in a patch to increase the potential number of dispersing cells.

This tradeoff between growth rate and biomass yield expresses a design conflict between short and long timescales. Short-acting processes favor rapid growth to outcompete neighbors.[236] Long-acting processes favor efficient yield to increase colonization of new patches. Different conditions cause different timescales to dominate (Section 5.9).[129,130]

Various timescale tradeoffs arise in the study of design. A few examples follow.

SHORT-TERM VERSUS LONG-TERM GAINS

A rapidly growing mutant lineage that outcompetes neighbors is similar to cancerous overgrowth.[135] The short-term reproductive benefit favors highly competitive design.

Over longer timescales, a growth-enhanced cancerous design loses to yield-efficient designs tuned to increase dispersal success over the full demographic cycle of local growth and subsequent dispersal.

The particular design that is ultimately favored depends on the balance between short-term gains by local competitive growth versus long-term gains by dispersal.

The following list summarizes a few of the timescale tradeoffs.

- Short-term local gain trades off against long-term global gain.

- Rate versus yield compares gains on short versus long timescales.

- Short-term reproductive gain within a patch trades off against long-term dispersal gain between patches.

- Short-term competitive gain between genomic subsets within an individual trades off against the individual's long-term success.

MULTILEVEL SELECTION

Natural selection can act at multiple levels. The different levels of selection associate with different timescales.[133,446]

For example, growth rate affects relative success against neighbors within a locally competing group. Growth rate competition within groups happens over shorter timescales.

Biomass yield affects the productivity of a lineage or group when competing against other groups for dispersal and colonization of new resource patches. Biomass yield competition between groups happens over longer timescales.

Selection can also happen on a lower level, between genomic subsets within an individual. That low-level genomic competition happens on a shorter timescale than competition between individuals.

For example, plasmids may compete within cells over short timescales while lowering the success of host cells over relatively longer timescales.

At a higher level, selection can act more slowly between species or clades. A clade can outcompete other clades ecologically over broad spatial and temporal scales.

A clade can also reproduce in the sense of splitting into descendant lineages. On a relatively slow timescale, we can consider species births and deaths, determining the relative dominance of genera in terms of numbers of species.

The designs that we see in nature depend on the balance between selective success acting at these different timescales. In general, the lower levels and faster timescales dominate because selection almost always happens more quickly and intensely at lower levels.[446]

Selection at a lower, faster level may trade off against selection at a higher, slower level.[168]

EVOLVABILITY: TRADEOFFS BETWEEN EXPLORATION AND EXPLOITATION

Organisms continually face new challenges. Long-term success demands the ability to adapt. Organisms gain by exploring alternative designs. Greater exploration speeds the rate of adaptation to new challenges and lowers the chance of lineage extinction.[311,436]

Mechanistically, higher mutation rates and more mixing of genomes to create new gene combinations increase exploration of alternative designs.[267,308] Other variation-generating mechanisms may also speed adaptation.[443]

Mechanisms that generate variation and improve exploration often reduce efficiency in exploiting current resources. Long-term evolvability trades off against short-term fitness.

For example, most mutations are deleterious in the short term. Greater mutation rate reduces short-term exploitation success. However, as the environment changes, a higher mutation rate may allow faster adaptation to changing conditions by exploring a wider range of alternative designs.

15.8   Bet-Hedging Tradeoffs

Bet-hedging is another form of exploration versus exploitation. Bet-hedging describes alternative states for a trait.[131,144,370]

The different states may be alternatives for a single individual. For example, an individual may feed in different locations to hedge its bets against settling in a lower quality patch.

Or the states may be alternatives for different individuals from the same genetically identical clonal lineage. For example, some cells may stay and exploit a good patch, whereas other cells may disperse.

The clone's overall success is the aggregate performance of its different bets on local versus distant gains. If the local patch disappears suddenly, the clone persists through its dispersers.

Bet-hedging may increase the success of a lineage by trading off the potential gains of exploration against the losses of reduced exploitation.

EXPLORATION ACROSS SPACE AND TIME

Exploration may occur spatially through motility and dispersal or temporally through persistence, dormancy, and sporulation.

- If an individual's motility randomly samples alternative locations, then its bet-hedging movements trade off against motility costs.

- Cells in a clone may split into dispersers versus nondispersers, dormant versus active cells, and spores versus nonspores. These polymorphisms hedge a genotype's bets across space and time.

- Cells within a clone may transiently enter and leave a persistent state that resists antibiotics.[421] For example, a temporary period of cellular reproductive quiescence protects against molecules that attack replicating DNA.

- Cells in a quiescent state trade a clone's growth for a hedge against attack.

- During starvation, a clone may hedge between rapid resumption of growth and resistance to attack.[282] Cells in an active state resume growth rapidly. Cells in a quiescent state may survive antibiotic attack. A clone hedges by maintaining some cells in alternative states, achieving both rapid growth and resistance.

- Traits such as quiescence or dispersal may arise inevitably from biophysics rather than from the biological forces of design. Interpreting design depends on tests of comparative hypotheses.

EXPLORATION OF ALTERNATIVE FOOD SOURCES

In the classic diauxic shift, cells first feed on a preferred carbon source until it is depleted. A lag period follows during which cells shift gene expression to prepare the catabolic system for feeding on a second, less preferred carbon source (p. 214).

- After the preferred source is depleted, some cells may switch to the secondary source. Other cells may remain activated for the primary source. The primary-activated cells do not grow in the absence of the primary source but remain ready to resume growth rapidly if more of the primary source arrives.

- Similarly, some cells may switch to the secondary source before the first source is depleted. Having some cells pre-activated for the secondary source reduces the clone's growth lag when the first source runs out.

- In each case, a clone adjusts more rapidly to changing availability of the two resources when it hedges between the alternative states. Hedging trades lower immediate growth for more rapid adjustment to changing conditions.

15.9 Control Tradeoffs

Control adjusts traits to internal and external conditions. Linking metabolic control to demography and life history poses a challenge for future work. Here, I briefly repeat a few key tradeoffs developed in Chapter 7.

- Error-correcting feedback trades off against the costs of control. *Error* is the difference between a trait's value and the target value for the current environment. Costs arise from measuring the error and adjusting traits.

- Robust error correction at the system level trades off against error-prone system components.

- Fast adjustment trades off against stability. Fast adjustment requires a strong push, which risks overshooting the target. Overshoot destabilizes a system, potentially causing it to move dangerously far from its viable range.

- Responsiveness trades off against homeostasis. The more easily a system can respond to change, the more easily it can be perturbed from its homeostatic setpoint.

15.10 Summary

This chapter emphasized broad conceptual issues and common tradeoffs. That general scope provides a sense of the potential challenges of design faced by organisms.

In practice, tradeoffs can be very specific. Consider the title: "Evolutionary trade-off between vocal tract and testes dimensions in howler monkeys."[91]

For microbial metabolism, nearly every biochemical and physiological detail could be associated with some tradeoff. In spite of that specificity, it is important to retain a broad sense of the main forces and tradeoffs that shape all of life.

Combining a wide conceptual approach with the specific understanding of natural history in each particular application emphasizes the many biological challenges and the many different tradeoffs faced by organisms.

The shifting dominance by different tradeoffs as conditions change raises a common difficulty in the study of design. As an example, I mentioned previously the complexity of analyzing tradeoffs between rate, yield, and antitoxin defense.

In a lab study that excludes interspecies toxin warfare, one may observe both rate and yield increasing. That joint increase may arise because rate trades off against antitoxin defense, and yield also trades off against defense. In the absence of attack, investment in defense declines, and both rate and yield increase.

One cannot test whether a tradeoff between two traits is important simply by measuring how the two traits change. Instead, one must develop comparative hypotheses about partial causation. For example, how does preventing attack alter the rate versus yield tradeoff in shaping design?

Comparative hypotheses do not solve all problems. But if observations tend to support a comparative prediction under a variety of conditions, then one may be on the right track. Further progress follows by adding additional partial causes to improve the success rate for predictions.

Finally, there remains a gap between the biochemical detail of prior chapters and the emphasis on fitness components in this chapter. Bridging that gap remains a central challenge in the study of design. The following chapters make a start.

16 Predictions: Overflow Metabolism

> Ever tried. Ever failed. No matter. Try again. Fail again. Fail
> better.
>
> —Samuel Beckett[30]

Every study that analyzes biological design should clearly state a comparative prediction. This book is about why we need such predictions and how to formulate predictions that reveal the causes of design.

This chapter develops predictions for overflow metabolism. Overflow happens when cells excrete intermediate metabolic products that contain usable free energy. Why do cells dump usable food?

The following chapter adds predictions for variable sugar usage, distributed electron flux, and alternative free energy stores in fluctuating environments.

These predictions provide a rough draft on which others may build or, for the more aggressive, a target to attack. No matter, as long as the result improves.

The first section restates the basic structure of comparative predictions. The second section recaps key observations about metabolic overflow. Those observations set the puzzles, highlight important concepts, and recommend overflow as a model for studying biological design.

The third section considers proteome limitation as a solution to the puzzle of overflow. Cell volume imposes a biophysical constraint on the number of proteins. Dumping usable food may happen because protein limitation imposes a tradeoff between growth rate and efficiency.

However, biophysical constraints alone mislead. Those constraints set broad boundaries on traits. Within those boundaries, changes in demography or in genetic variability between competitors may greatly alter design. This section presents observations and new predictions to clarify how natural history and biophysical constraints interact.

The fourth section evaluates limited membrane space as an alternative constraining force. More membrane transporters for food reduce the membrane space for electron transport and oxidative phosphorylation. Thus, faster food uptake and metabolic flux may reduce aerobic respiration, causing cells to overflow intermediate metabolic products.

Combining proteome and membrane constraints gives a sense of the multidimensional biophysical boundaries that limit possible metabolic designs. Comparative predictions evaluate the various forces that alter traits within that space of possible designs.

The fifth section describes a single genetic change that broadly alters metabolic flux. In fission yeast, fast versus slow pyruvate kinase activity shifts cells from glycolytic overflow and reduced aerobic respiration to limited overflow and nearly complete aerobic processing.

Natural populations contain both genetic variants. Apparently, some environmental conditions favor fast metabolic flux, rapid growth, and reduced yield associated with overflow excretion. Other environmental conditions seemingly favor the opposite.

The slow allele associates with greater tolerance of oxidative stress. Perhaps environments vary in oxidative challenge, explaining the different metabolic designs. Alternatively, other environmental challenges may dominate.

16.1   Comparative Predictions and Partial Causes

Restating briefly, a comparative prediction may be expressed by $P \rightarrow F \rightarrow T$. As the environmental parameter, P, increases, the trait, T, increases, mediated by the force of design, F. *Environmental* means any causal factor, which could be external or internal to an organism (Chapter 3).

We may have decreasing relations given by $\dashv$ instead of $\rightarrow$, leading to four basic expressions for partial causes.

The causes are partial because other causes may also influence the trait. Ultimately, all partial causes combine to determine the overall cause. However, we almost never know all partial causes.

I focus on partial causes because each pathway expresses how a force of design partially shapes a trait. We test a prediction by studying many instances of a change in P under many distinct circumstances.

Ideally, the different circumstances randomize the other partial causes sufficiently so that the focal pathway has significant effect. If that randomization succeeds, then the predicted direction of change should happen more often than not.

Obviously, a lot can go wrong. The best we can do is mitigate confounding factors that we identify and openly accept the limitations for those factors that we cannot identify.

Mitigating known factors often means identifying other pathways of partial causation. If there is a second causal parameter, $\tilde{P}$, we might predict that when $\tilde{P}$ is high, we expect $P \rightarrow T$, and when $\tilde{P}$ is low, we expect $P \dashv T$. We must also consider common ancestry, a frequent confounding cause in evolutionary studies.[173]

The benefits of comparison in causal analysis are well known. However, in the study of design, it is often difficult to obtain meaningful comparative data. It is particularly hard to get the right scale over which parameters change and organisms respond such that the comparative tests reveal design.

The fact that it is hard does not alter the need to think in this way. Developing comparative predictions starts the process.

The individual statements of partial cause form the building blocks for simple empirical tests and for understanding more complex combinations of causal interactions.

16.2   Background

The faster a cell takes up simple sugars, the more it tends to excrete glycolytic products. The glycolytic products contain almost all of the potentially usable free energy (Fig. 12.2). With more rapid digestion, the catabolic cascade seemingly overflows (Section 12.2).

THE IMPORTANCE OF A GOOD PUZZLE

We do not expect cells to dump potentially usable food unless strong forces or powerful constraints favor such losses. Several aspects of metabolism likely play a role. Thus, overflow metabolism is a particularly good puzzle because its solution will likely reveal many facets of design.

In a typical case, overflow metabolism occurs in cells capable of aerobic respiration. Such cells can pass a simple sugar through glycolysis, the TCA cycle, electron transport, and oxidative phosphorylation (Fig. 16.1). Glycolysis yields only a small amount of ATP and usable free energy.

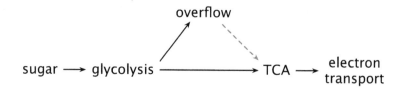

Figure 16.1 Catabolic flux and glycolytic overflow in cells capable of aerobic respiration. As sugar intake and glycolytic flux rise, cells may excrete glycolytic products. In *E. coli* at low sugar intake rate, all flux passes through the TCA cycle and electron transport, with no glycolytic overflow.[28] As intake rises, post-glycolytic flux does not keep up and excess glycolytic flux overflows as excreted acetate. In *S. cerevisiae*, rapid sugar intake associates with excreting post-glycolytic flux as ethanol. After consuming the sugar, yeast cells may shift to catabolizing the ethanol through the TCA cycle, electron transport, and oxidative phosphorylation (dashed arrow).[49] This diagram is a copy of Fig. 12.4, repeated here for convenience.

Nearly all of the ATP and usable free energy that the cell could extract comes from catabolizing the glycolytic products. Why not process the glycolytic products through the TCA cycle and electron transport?

The following subsections list possible solutions mentioned in prior chapters. After that review, I develop some comparative predictions.

THERMODYNAMIC PRODUCT INHIBITION

Net forward flux halts when reaction products accumulate. In cells capable of aerobic respiration, glycolysis would halt if glycolytic products are made more quickly than they can be processed through the TCA cycle and electron transport. Cells may excrete glycolytic products to relieve that product inhibition.

REDOX IMBALANCE

What might cause post-glycolytic flux limitation? The TCA cycle transforms NAD^+ to NADH. An excess NADH–NAD^+ disequilibrium impedes flux through the TCA cycle by product inhibition. NADH is a strong reducing agent. To maintain TCA flux, cells must dissipate the redox imbalance caused by an excess NADH–NAD^+ disequilibrium.

MEMBRANE SPACE LIMITATION

What might limit the rate of dissipating the NADH–NAD^+ disequilibrium? The NADH–NAD^+ disequilibrium drives electron transport, which drives oxidative phosphorylation. When electron transport flux does not keep up with NADH production in the TCA cycle, the NADH–NAD$^+$ disequilibrium builds up and reduces TCA flux.

Rapid glycolytic flux requires additional membrane transporters for food uptake. Those additional transporters may limit the membrane space available for electron transport. Limited electron transport requires excretion of glycolytic products to reduce TCA flux and the NADH–NAD$^+$ disequilibrium.

ELEMENTAL LIMITATION

Alternatively, elemental limitation may slow post-glycolytic flux. Scarcity of oxygen or other final electron acceptors reduces electron transport flux. Scarcity of certain metals or other elements may limit particular catalysts or electron-transport cytochromes in ways that reduce TCA or electron-transport flux more strongly than glycolytic flux.

PROTEOME LIMITATION

Constrained cellular space and resources limit total protein abundance, limiting the protein catalysts that drive flux. Limited catalysts impose flux tradeoffs between alternative pathways.

As sugar uptake rate and glycolytic flux increase, cells often grow faster. That faster growth requires more catalysts to drive anabolic processes. The extra catalysts for growth must be balanced against fewer catalysts for other processes.

If glycolytic flux produces ATP at a sufficient rate to sustain fast growth, then cells can limit flux through the TCA cycle, electron transport, and oxidative phosphorylation. The proteome fraction that might have been used to drive those post-glycolytic processes can instead be used for anabolic processes.[28]

With reduced TCA flux, cells must excrete post-glycolytic products, leading to overflow metabolism.

REPRODUCTIVE RATE VERSUS YIELD

Overflow metabolism associates with high reproductive growth rate and low reproductive yield per unit of food. Cells often grow faster by excreting glycolytic products but, by discarding usable free energy, their reproductive yield declines.

Design forces shift the balance between rate and yield. For example, more competition between distinct genotypes favors greater rate and lower yield, causing a rise in overflow. Changes in dispersal and demography also tend to shift the favored rate-yield balance.

THERMODYNAMIC RATE VERSUS YIELD

In a catabolic cascade, the more free energy yield that is captured and stored for later use, the less the total free energy will change. Less free energy change means lower thermodynamic driving force and slower flux. Thus, the flux rate of a catabolic cascade trades off against the captured free energy yield (Section 11.1).

In the literature, the thermodynamic rate versus yield tradeoff is often discussed implicitly or directly as the cause of overflow metabolism and the reproductive growth rate versus yield tradeoff. Thermodynamics inevitably plays a key role. But we must consider carefully the potential connections between thermodynamics, metabolic flux rates, overflow metabolism, and reproductive tradeoffs.

First, thermodynamic driving force influences flux rate but does not determine it. Resistance also strongly influences flux. Resistance may be altered by catalysts and by the spatial movement of molecules. Changes in proteome allocation for catalysis may influence reproductive tradeoffs between growth rate and biomass yield. Other resource reallocations may also affect the balance between growth and yield fitness components.

Second, to analyze overflow metabolism, we compare the excretion of a post-glycolytic product with the processing of that product through respiration. That comparison contrasts two alternative reaction cascades. Thermodynamic constraints influence reaction cascades, but tradeoffs between flux rate and free energy yield capture may be complex when comparing alternative cascades.

TRADEOFFS WITH OTHER FITNESS COMPONENTS

Overflow typically associates with increased growth rate and decreased reproductive yield. Thus, conditions that favor rapid growth rate may favor overflow metabolism.

Emphasis on rate and yield highlights two fitness components. However, other fitness components may also be important. For example, greater aerobic respiration associates with enhanced tolerance of oxidative stress. Many environmental factors cause oxidative stress, including chemical warfare between microbial species.

Variations in oxidative stress and survival may tip metabolism more strongly toward or away from glycolytic overflow. For example, an increase in oxidative stress favors aerobic respiration, causing declines in glycolytic excretion and growth rate (Section 16.5).

The extra reproductive yield associated with efficient respiration may be lost to increased death through oxidative stress. The net effect of rising oxidative stress could be an overall decline in rate, yield, and overflow. Tradeoffs always depend on broader context.

CHALLENGES OF ANALYZING DESIGN

Why does cellular metabolism overflow? That puzzle poses the question of design: What makes the organism as we see it?

However, that question is too hard. Many different forces have acted over many different timescales. We cannot evaluate all of those forces.

We must ask different questions. How does a particular force influence what we currently see in the context of the other forces currently operating? How many such partial causes can we identify? How do those partial causes interact?

We may think of those questions as a local analysis. How can we understand what makes the organism as we see it relative to nearby alternatives that might occur?

Comparative predictions link partial causes to nearby alternatives. Those predictions provide the pieces that must be fit together to solve a particular puzzle of design.

The following sections extend three prior studies, considering a wider range of constraints and fitness components. I highlight those constraint and design forces through comparative predictions.

16.3   Proteome Limitation

Overflow metabolism occurs in *E. coli*. At sugar uptake rates below a critical threshold, aerobically catabolizing cells process most sugar through glycolysis, the TCA cycle, and oxidative phosphorylation. Faster sugar uptake associates with faster growth.[28]

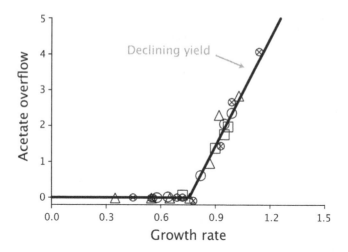

Figure 16.2 Overflow metabolism in *E. coli*. Symbols show various strains grown under different conditions. Growth rate is measured per hour, with doubling per hour at $\log(2) = 0.69\,h^{-1}$. Acetate overflow is given as excretion rate per hour per unit of bacterial biomass. With increasing overflow, biomass yield per gram of sugar taken up declines. See Section 12.2. From Figure 1 of Basan et al.[28]

Above the uptake-rate threshold, cells transform the additional sugar that passes through glycolysis into excreted acetate. Faster sugar uptake associates with more excreted acetate and faster growth (Fig. 16.2).

Proteome limitation may explain the observed patterns of sugar uptake and overflow acetate excretion[28,279,313] (Fig. 16.3). The cell must allocate its limited number of proteins between aerobic respiration to make ATP and anabolic construction to make new cells.

When growth is slow and protein is not limiting, then allocating protein to both glycolysis and post-glycolytic pathways extracts the greatest value from food.

As growth rate increases, demand for protein rises. Cells increasingly favor enhanced efficiency per unit of protein allocated to different functions.

Glycolytic fermentation and acetate excretion apparently have a higher efficiency of ATP production per unit protein than does the TCA cycle. Greater growth rate and anabolic demand may favor cells to enhance the more protein-efficient glycolytic fermentation pathway relative to the TCA pathway.

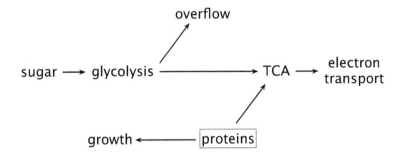

Figure 16.3 Overflow metabolism arises from proteome limitation. At high sugar uptake rate, limited protein imposes a tradeoff. Additional protein may be allocated either to TCA catalysis or to anabolic cellular processes of growth and reproduction. Augmented TCA catalysis reduces glycolytic overflow and increases catabolic efficiency but limits growth and reproduction. Alternatively, protein allocated to augmented growth reduces TCA catalysis, forcing overflow of some glycolytic products.[28,279,313]

Basan et al.'s[28] experiments support proteome limitation as a key constraining force. The proteomic constraint explains the changing physiological response of various *E. coli* strains to changing laboratory conditions.

Basan et al. conclude that overflow metabolism in diverse prokaryotic and eukaryotic cells likely follows from proteome limitation. In other words, the proteomic constraining force may be sufficient to solve the widespread puzzle of overflow metabolism.

However, most experiments have focused on the physiological response of a few lab strains to a limited set of conditions. Different environmental conditions or different genotypes or different species may push up against different constraining forces.

Additionally, various forces of design may alter the favored balance between efficiency in terms of growth rate and efficiency in terms of biomass yield per unit of sugar taken up. Greater weighting of efficiency in biomass yield and less weighting of growth rate may favor increased protein allocation to catalyzing the TCA cycle, pushing cells toward reduced overflow.

Forces of design may also change the limitations imposed by constraints. For example, selection may modulate speed versus efficiency tradeoffs in electron transport by altering the properties of membranes and cytochromes (Section 16.4). Changes in electron transport modify throughput bottlenecks in catabolism and shift the relative valuation of different proteome allocation strategies.

Comparative predictions highlight the various forces of constraint and design.

PROTEOME CONSTRAINT

The broad comparative hypothesis is

proteome limitation → marginal benefit fermentation → overflow.

As protein becomes more limiting, fermentation's marginal benefit rises because of its greater ATP productivity per unit protein. Greater marginal benefit for fermentation favors more overflow. Protein limitation can be increased by the following methods to test this prediction.[28]

- Overexpressing a particular protein crowds out other proteins.

- Reducing mRNA translation slows protein production.

- Increasing sugar uptake and growth raises proteomic demand.

- Dissipating free energy raises catabolic and proteomic demand.

In each case, greater protein limitation predicts increased overflow metabolism. The physiological response of *E. coli* lab strains grown in batch culture supports these predictions.[28] The observed quantitative changes matched predictions from a simple model of proteome limitation, exceeding this qualitative summary.

ALTERNATIVE CONSTRAINTS

In Basan et al.'s[28] laboratory environment, proteome limitation dominated the expression of overflow metabolism. Many natural environments may impose the same dominant constraining force. However, some environments may push cells up against different constraints.

Earlier in this section, I described alternative constraints, such as membrane space limitation or elemental limitation. Those alternative constraints may cause post-glycolytic thermodynamic inhibition, leading to the broad comparative hypothesis

post-glycolytic inhibition ⊣ marginal benefit respiration ⊣ overflow.

The buildup of thermodynamic product inhibition within the TCA cycle or oxidative phosphorylation reduces the benefit of additional respiratory flux. The reduced benefit of passing additional glycolytic products through the TCA cycle favors increased overflow of those products through fermentation pathways.

Particular causes of thermodynamic inhibition suggest specific predictions. Causes of inhibition include

- Excess NADH–NAD$^+$ disequilibrium and redox imbalance.

- Flux limits through electron transport or oxidative phosphorylation.

- Elemental constraints that limit post-glycolytic catabolic enzymes.

- Limited oxygen or other post-glycolytic final electron acceptors.

In addition to these particular causes of thermodynamic inhibition, any necessary reaction in the TCA cycle and respiratory cascade may impose a flux limit and post-glycolytic inhibition. In flux = force / resistance, limited flux in a particular reaction may arise by restricted thermodynamic driving force or by increased resistance from proteome constraints on enzyme availability.

LAB STUDIES AND THE FORCES OF CONSTRAINT

Lab studies often emphasize the forces of constraint. Two reasons favor that emphasis.

First, such constraints clearly play an important role in shaping design. For example, if the total protein content of a cell is constrained, then more proteins allocated to one function leave fewer proteins available for other functions.

Second, manipulating the lab environment pushes cells up against various constraints. For example, limitation of a particular nutrient enhances certain constraints. Excess of another nutrient shifts the dominating constraint to a different factor.

Various stresses can be manipulated. Genetic knockouts may force compensatory responses that push against previously latent constraints. Enhanced gene expression or novel genes may find the physiological boundaries of yet other constraints.

FORCES OF DESIGN MODULATE TRAITS

The forces of design operate over longer evolutionary timescales than the physiological responses that reveal short-term forces of constraint. Consequently, fewer studies directly analyze environmental changes that influence the forces of design. Nonetheless, such forces play a key role.

This section focuses on proteome limitation, a particular constraining force that may influence overflow metabolism. How can we relate the forces of design to proteomic constraints and overflow metabolism? In general, how do the forces of design modulate traits within the context of constraining forces?

Under proteome limitation, more catalysts for anabolic growth mean fewer catalysts for the TCA cycle. The constraining force of limited protein causes a tradeoff between growth rate and the yield-enhancing efficiency of post-glycolytic catabolism.

The constraining force of proteome limitation by itself does not set the relative allocation to faster growth versus more efficient yield. Instead, the forces of design tune trait values within the context of that constraining force.

If environmental conditions change such that the forces of design favor faster growth and lower yield, then the relative proteome allocation will tend to shift more strongly toward anabolic processes and away from post-glycolytic catabolism, increasing the tendency toward overflow of post-glycolytic products.

What about the constancy of the constraints? In other words, can design forces drive evolutionary change that modulates the constraints?

For example, limits on proteome size arise partly from traits that evolve. Cell size may change, altering the space available for proteins. Cells can trade off making more proteins against other cellular functions. Often, evolution can alter the relative importance of different tradeoffs.

It is easy to identify those broad conceptual issues. But can we understand how actual forces shape the design of observed traits?

As always, there is no simple way. Comparative predictions provide the best path forward. To suggest how one might start, I briefly review three general aspects in the context of proteome limitation. For each, I sketch the preliminary comparative hypotheses that follow.

Evolutionary constraints.—Many lab studies push cells up against physiological constraints. How important are those physiological constraints for the evolutionary design of traits?

On the one hand, there must be limits to how fast cells can take up sugar or create membrane ion gradients that drive ATP production. On the other hand, evolutionary changes may modulate physiological limits.

For example, modified receptors may alter limits on uptake rate. Modified catalysts may alter flux limits. Evolutionary changes may increase or decrease the constraining rate limits.

In other words, evolutionary change can modulate physiological constraints. But there must be some constraints on evolutionary change. Those evolutionary constraints play a key role in biological design.

Evolutionary constraints can be difficult to observe. Cheng et al.'s[65] experimental evolution study of *E. coli* provides some clues (p. 175).

The independently evolved lines increased growth rate by approximately 50%. Overflow acetate excretion and biomass yield also evolved (Figs. 16.4 and 16.5).

Acetate overflow increased with glucose uptake and declined with biomass yield, as expected from biochemistry. Interestingly, the evolved increase in growth rate did not associate strongly with changes in uptake, yield, or overflow. Instead, for a given evolved growth rate, the yield, y, and the uptake rate, q, varied widely along the inverse relation $y \propto 1/q$ set by the definitions of the variables (Fig. 12.5 and eqn 12.1).

The evolutionary independence of growth relative to the other metabolic variables contrasts with the strong physiological associations observed in other studies. Physiologically, growth rises with glucose uptake rate. Above a growth rate threshold, acetate excretion rises linearly with uptake and growth rates (Fig. 16.2).

The difference between evolutionary and physiological tradeoffs suggests that evolutionary response can modulate physiological constraints. Thus, one cannot draw strong conclusions about design based on observed physiological constraints. Instead, one must evaluate limits on the potential paths of evolutionary change caused by constraining forces.

Put another way, physiological limits certainly exist and shape design. However, observed physiological tradeoffs are not the same as absolute biophysical limits.

How should we analyze design, given the complexity of tradeoffs arising from physiological and evolutionary constraints? In my prior

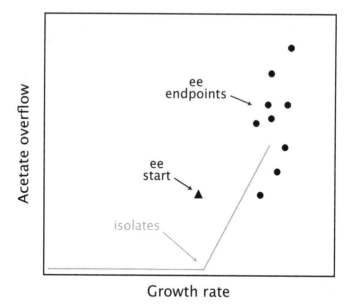

Figure 16.4 Experimental evolution shows how the forces of design modulate physiological constraints. Various isolates of *E. coli* fall along the gray line when grown under different physiological conditions (Fig. 16.2). Basan et al.[28] suggested that the close fit of the isolates to that line arises by a growth rate versus biomass yield tradeoff imposed by the physiological constraint of proteome limitation. Cheng et al.'s[65] experimental evolution study shows that design forces can move traits off the constraint line. When the experimental evolution (ee) starting strain was subjected to natural selection favoring faster growth, the independently evolved lines changed to the endpoints. For a given growth rate, greater acetate overflow corresponds to higher glucose uptake rate and lower biomass yield (eqn 12.1). Redrawn from Fig. 1B of Cheng et al.[65]

discussion of Cheng et al.'s[65] study, I proposed a new interpretation of the observed relations between growth rate, yield, sugar uptake rate, and acetate excretion rate (p. 175).

The experiment imposed selection in a way that limits time for growth rather than limits sugar for growth. Because the imposed design force pushes strongly on biomass produced per unit time (growth rate) and weakly on biomass produced per unit sugar (yield), the yield is effectively a neutral trait.

Under these conditions, natural selection favors increased growth rate independently of the consequences for yield. Thus, the various evolved

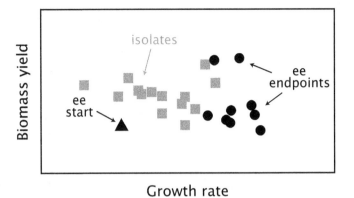

Figure 16.5 Experimental evolution increases growth rate, with biomass yield changing in an apparently neutral and uncorrelated way. By contrast, isolates grown under different conditions follow a rate versus yield tradeoff, as reported in Basan et al.[28] and illustrated in Fig. 16.2. Redrawn from Fig. 1A of Cheng et al.[65]

lineages explored alternative physiological mechanisms to achieve similar levels of increased growth. A given level of growth associated with different levels of biomass yield, y, and uptake rate, q, with values along the curve $y \propto 1/q$ set by the definition of those quantities.

As always, tradeoffs depend on context. The common tendency to apply tradeoffs without considering context leads to a poor understanding of design. This particular study considered the physiological tradeoffs revealed by previous data but did not emphasize the special evolutionary context of the experimental setup.

Comparative predictions focus the conceptual issues and provide a way forward. For Cheng et al.'s experiment, the apparent neutrality of yield among the experimentally evolved lines calls attention to the tradeoff between growth rate and yield. For example,

sugar limitation → marginal benefit yield → rate-yield tradeoff. (16.1)

As sugar becomes more limiting, the benefit from enhanced biomass yield per unit sugar intake increases. The greater the marginal benefit of yield, the more strongly gains in growth rate trade off against the loss from reduced yield.

In this particular experimental design, each bout of growth begins with abundant sugar. Cells grow to approximately mid-exponential phase and are then passaged to a renewed environment, never approaching stationary phase.[65,219]

Abundant sugar means little sugar limitation, low benefits of yield, and a weak tradeoff between growth rate and yield. We can express the same idea in terms of time limitation rather than sugar limitation,

time limitation ⊣ marginal benefit yield → rate-yield tradeoff.   (16.2)

As time becomes more limiting, the benefit declines for using sugar efficiently to enhance yield. With limited time, growing as quickly as possible independently of yield often provides the highest fitness.

These predictions about sugar and time limitation may be tested by varying both in a controlled way. How finely can the forces of design tune growth rate and yield? What biochemical and physiological processes impose the most important forces of constraint? What environmental factors push cells up against different constraints and result in different tradeoffs in evolutionary tuning?

Returning to the experimental results of this particular study, yield varied inversely with glucose uptake rate among the evolved lines. In other words, the full aerobic catabolic cascade did not keep up with increasing sugar uptake. What might cause that tradeoff between uptake rate and yield? Mechanistically, is there a particularly important bottleneck in the catabolic cascade?

Cheng et al.'s[65] metabolic modeling pointed to NADH-NAD$^+$ disequilibrium, a catabolic bottleneck that I have mentioned several times. High TCA cycle flux creates a strong NADH-NAD$^+$ disequilibrium. Failure to dissipate that disequilibrium sufficiently rapidly causes the catabolic cascade to slow and intermediate products to accumulate. Acetate overflow and low yield provide one solution.

Typically, much of the NADH-NAD$^+$ disequilibrium dissipates by driving electron transport and oxidative phosphorylation. The disequilibrium may build if electron transport flux does not keep up with TCA flux. One solution to an excess NADH-NAD$^+$ disequilibrium, emphasized by Cheng et al., is to dissipate the disequilibrium more rapidly.

Electron transport and oxidative phosphorylation can increase the rate at which they convert NADH to NAD$^+$ by using that driving force less efficiently. In other words, a lower number of ADPs converted to ATPs per NADH to NAD$^+$ conversion means less free energy capture, greater thermodynamic driving force, faster dissipation of the NADH–NAD$^+$ disequilibrium, and more rapid flux through electron transport.

Lower ATP yield and greater flux rate may associate with less efficient cytochromes in electron transport or a leaky membrane that partially dissipates the proton gradient used to drive oxidative phosphorylation.

Those mechanisms would lower the yield for flux through the respiration pathway. However, the reduced yield for less efficient respiration may be small relative to the reduced yield for increased acetate overflow. Comparing yield between different pathways takes us back to the role of proteome limitation, which may affect pathways differently.

Preliminary data suggest that cells may be able to use less efficient cytochromes to increase flux and reduce yield.[472] It would be interesting to test comparative hypotheses. Conditions that strongly favor both fast growth rate and high yield may favor an evolutionary response that maintains respiratory flux rather than increases overflow excretion, with faster and less efficient cytochromes or a leakier membrane with respect to the proton motive force that drives oxidative phosphorylation.

Theory of rate versus yield.—This section began with proteome limitation as a possible cause of overflow metabolism. Proteome limitation imposes a physiological force of constraint. Presumably, evolutionary forces of design interact with that constraining physiological force.

The experimental evolution study of Cheng et al. provides insight into interactions between forces of design and constraint. In my interpretation, their experimental setup created design forces that acted strongly on growth rate and weakly on biomass yield.

With only weak design forces acting on yield, any constraining forces that may couple rate and yield became relatively unimportant. Thus, yield and acetate overflow drifted in an approximately neutral way relative to evolutionary changes in growth rate.

That tentative interpretation focuses attention on the design forces that shape growth rate and biomass yield. Alternative conditions may impose strong design forces on both rate and yield, leading to different predicted outcomes for evolutionary response. To make further progress, we must consider comparative predictions.

I presented many rate-yield predictions in earlier chapters. I mention a few here to illustrate application. In Section 4.1, the prediction

$$\text{mixing} \dashv \text{relatedness} \dashv \text{rate}$$

states that greater genetic mixing decreases the relatedness among competitors. Lower relatedness favors higher growth rate because individuals gain by outgrowing genetically different competitors. That general prediction from kin selection theory refines the predicted influence of time limitation on growth rate (eqn 16.2).

For example, when sugar is limited, relaxing time limitation favors slower growth and greater yield. However, high genetic mixing and low relatedness favor faster growth and lower yield.

With two distinct causes that have different predicted effects, one can study a broader range of design forces that may tune growth rate and yield. Extensions to the experimental setup of Cheng et al. provide an opportunity to manipulate those forces and test comparative hypotheses.

Section 4.1 also presented the prediction

$$\text{patch lifespan} \rightarrow \text{marginal yield} \dashv \text{rate.}$$

Here, we may consider patch lifespan as a natural cause of time constraint, in which a longer patch lifespan with fixed resources relaxes time limitation for growth. We then recover a variant of the prediction in eqn 16.2 that links the laboratory experiments to natural history.

Section 10.3 lists several predictions about growth rate. For example,

$$\text{attack} \rightarrow \text{marginal defense benefit} \dashv \text{rate}$$

states that increased attack by neighbors raises the marginal benefit of investment in defense. Greater investment in defense may reduce resources for growth rate.

In Cheng et al.'s setup, what sort of metabolic changes might evolve in response to intense attack? Would reduced overflow and greater post-glycolytic flux arise in order to capture more free energy that could be used for warfare? Or would intense warfare require greater free energy flux rate at the expense of lower metabolic efficiency? Or perhaps warfare would reduce expected survival time in the local resource patch, favoring rapid growth and dispersal to escape attack.

How would investment in warfare trade off against growth rate and biomass yield? Would biophysical constraints of membranes become important, with tradeoffs between sensitivity to attack, nutrient uptake, and electron transport leakiness? New theory that clarified the comparative predictions would be useful, along with direct empirical tests.

With regard to proteome limitation, we may consider environmental factors such as time constraints and sugar constraints and warfare. How does varying those factors affect physiological changes within particular strains in relation to proteome limitation? How does evolutionary response tune allocations to alternative functions in the context of constraining protein limitation?

What do we learn by comparing physiological limits within strains with the contours of trait changes in short-term evolutionary responses? Perhaps evolutionary response reveals more clearly the strong biophysical limits within the broad genetic architecture of a strain, whereas physiological limits may also include constraints of the specific regulatory control responses of that genotype.

A lab study favoring high yield.—Typical lab protocols for experimental evolution use large mixed populations in which cells compete globally for resources. Such mixing favors growth rate over yield.

Bachmann et al.'s[21] novel experimental protocol favors yield over rate. They created many isolated populations, each typically founded by a single cell. Each isolated population consists of a water-based droplet of growth medium separated from other droplets by an oil phase.

An isolated clone in a droplet grows without competition from other clones. After a period of growth, the clonal outputs are mixed into a large population. A new round of isolated droplets then forms by colonization from the mixed population.

If the droplet phase uses up most local resources, then fast initial growth provides no benefit. Instead, over the full demographic cycle, those clones that use food with the greatest reproductive efficiency will yield the greatest number of progeny. Design forces favor high yield.[401]

This study analyzed the evolutionary tuning of growth rate versus yield by competing two strains of *Lactococcus lactis*. The wild-type bacterium transforms the glycolytic product pyruvate into excreted lactate. A mutant strain without lactose dehydrogenase transforms pyruvate into excreted acetate (Fig. 16.6).

Lactate production yields 2 ATP per glucose molecule taken up. Acetate production yields 3 ATP. Comparing pathways, the lower ATP yield for lactate associates with faster growth and lower biomass yield per gram of glucose.

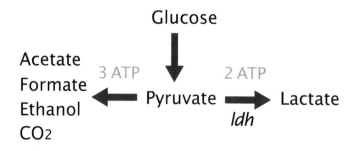

Figure 16.6   Alternative fermentation pathways in the anaerobic bacterium *Lactococcus lactis*. Lactate production associates with lower ATP yield, lower biomass yield, and faster growth rate compared with the alternative fermentation pathway leading to acetate, ethanol, or other products. A knockout of lactose dehydrogenase (*ldh*) creates a strain that follows the low rate and high yield pathway. Redrawn from Fig. 1C of Bachmann et al.[21]

In a mixed environment with direct competition between strains, the lactate-producing genotype won because it grows faster. For the water droplet in oil demography, the acetate-producing mutant won because it has a higher yield.

High-rate and low-yield lactate excretion is not typically described as overflow metabolism. But conceptually it is the same as overflow. In both cases, cells grow faster by dissipating usable free energy and have a higher ATP production rate at the expense of reduced ATP yield.

This particular experiment competed two alternative genotypes. Competition between qualitatively distinct alternatives does not provide insight into the quantitative tuning of metabolism in response to varying design forces. But the experiment does suggest new protocols for studying the evolutionary tuning of metabolism.

My previous summary of Cheng et al.'s[65] experimental evolution study of *E. coli* described how an increased design force favoring high growth rate altered metabolism. In a study that favored high yield, such as the oil emulsion protocol, what evolutionary changes in outcome and mechanism would occur for growth rate, ATP and biomass yield, post-glycolytic excretion, and TCA cycle flux?

If it is possible to create a demography that imposes strong forces favoring both fast growth and efficient yield, how would those forces tune electron transport and oxidative phosphorylation with respect to cytochrome efficiency, membrane leakiness for proton gradients, dissipation of the NADH–NAD$^+$ disequilibrium, and ATP production efficiency? What environmental changes would push cells up against different limits and impose different dominant tradeoffs?

Rephrasing those questions as comparative predictions would sharpen thought, highlight open conceptual and empirical problems, and lead to testable hypotheses.

SUMMARY

Lab studies often push cells up against constraining physiological and biophysical forces. Basan et al.'s[28] detailed analysis of proteome limitation provides a good example.

An observed limiting factor raises two questions. Is that the dominant force of constraint that sets the primary tradeoff shaping design? If so, how do varying forces of design modulate traits in the context of that dominant tradeoff?

Speculating, organisms likely live within a high-dimensional space bounded by many constraining forces. Various changed conditions push individuals up against different physiological and biophysical limits.

Each limit enhances the importance of particular tradeoffs and lessens the importance of other tradeoffs. Design forces move organisms within that bounded space over the short term and can sometimes alter the bounds over the long term.

With regard to overflow metabolism, constraints other than proteome limitation and design forces other than growth rate versus yield may often be important.

The next section considers membrane space limitation as an alternative constraint. The following section discusses growth rate versus survival against oxidative stress as an alternative design force.

16.4   Membrane Space Limitation

Different aspects of efficiency influence overflow metabolism. In a clonal population structure with limited resources, ATP yield per gram of food dominates. The greatest yield is achieved by aerobic respiration.

More competition between genotypes favors faster growth rate. ATP yield per unit time dominates. The best way to increase the ATP production rate depends on how various factors limit rate.

Proteome limitation may occur as growth rate increases.[279] For ATP production rate per unit protein, glycolysis is more efficient than the TCA cycle.[28] Excreting glycolytic products rather than passing them through the TCA cycle raises efficiency.

Alternatively, membrane surface area may limit nutrient uptake and electron transport, constraining growth rate. Membrane limitation favors greater rate efficiency of ATP production per unit surface area.[399,472]

This section develops the membrane limitation hypothesis. I then contrast the membrane limitation and proteome limitation arguments. Each limit favors particular aspects of efficiency and catabolic design.

OVERVIEW

Nutrient uptake may compete with electron transport for membrane space. Increasing nutrient uptake requires more membrane-bound transporters. Greater respiratory flux uses more membrane space for cytochromes and ATP synthase.

Electron transport provides efficient ATP yield per food molecule but a low ATP production rate per membrane area. Calculations for *E. coli* suggest:[399] "Acetate fermentation can ... produce the same amount of ATP [per second] as respiration using roughly a quarter of the membrane space, at the expense of a much lower ATP yield per glucose."

If membrane area becomes limiting at higher catabolic throughput, cells may enhance ATP production rate by relatively greater allocation of membrane space to nutrient uptake and less allocation to electron transport. Reduced electron transport imposes a TCA cycle bottleneck.

The TCA cycle increases the $NADH-NAD^+$ disequilibrium. Electron transport dissipates that disequilibrium to drive ATP production. Limited electron transport raises the net $NADH-NAD^+$ disequilibrium, favoring glycolytic overflow and reduced TCA flux.[399,472]

PREDICTIONS

Comparatively,

uptake & growth → membrane competition → overflow.

Rising nutrient uptake and growth increase competition for membrane space, which increases glycolytic overflow. *Escherichia coli* data support the association between membrane crowding and overflow (Fig. 16.7).

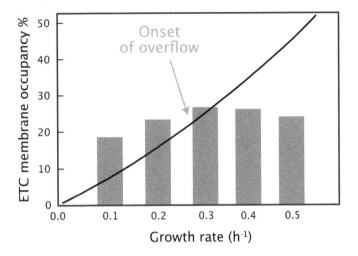

Figure 16.7 Percentage of the *E. coli* inner membrane occupied by electron transport chain (ETC) complexes at different growth rates. Observed occupancy is shown relative to the estimated maximum protein content on the membrane. As growth rate increases from zero, greater glucose uptake and catabolic flux raise membrane demand for respiratory proteins. The curve shows the theoretical minimum ETC occupancy needed to maintain full aerobic respiration with no acetate overflow. For these data, the observed acetate overflow excretion begins at $0.27\,h^{-1}$. As growth rate continues to increase, acetate overflow rises and observed ETC occupancy declines with the decrease in TCA flux and electron transport. At higher growth rates, the membrane fraction devoted to glucose transporters may rise, crowding out ETC components. However, that space competition has not been demonstrated empirically. Redrawn from Fig. 2 of Szenk et al.[399] Original data from Valgepa et al.[418,419]

Additional comparative predictions provide further insight.[472] For example,

membrane crowding → membrane competition → overflow.

Greater membrane crowding increases competition for membrane space, which increases overflow metabolism.

Membrane crowding may be increased experimentally by overexpressing a membrane-bound protein. In nature, environmental challenges such as a rising deficiency in iron or other nutrients may require more surface receptors, which could increase membrane crowding and overflow metabolism.

In prokaryotes, the surface area to volume (S/V) ratio determines the membrane area relative to the metabolic processes driving demand.

The S/V ratio in rod-shaped *E. coli* depends primarily on cellular radius. Length has little effect. Cells grow primarily by elongation, maintaining a roughly constant S/V ratio during cellular growth and division.[31]

Starved *E. coli* cells become smaller and more spherical, increasing their S/V ratio.[224] Other causes of changing S/V ratio may occur. If membrane limitation partially causes overflow metabolism, then

S/V ratio → available membrane ⊣ overflow metabolism.

As always, one must consider other partial causes. In this case, the need for scarce resources may demand more surface receptors and a greater S/V ratio. If the extra membrane relative to volume is taken up by specialized receptors, then the available membrane for respiration may not increase.

For *E. coli*, Zhuang et al.[472] mention several interesting predictions with regard to the three cytochrome oxidases of electron transport, Cyo, Cyd-I, and Cyd-II. These predictions emphasize efficiency in ATP production rate per unit membrane area.

Cyd-II relative to Cyo has greater respiratory throughput and lower ATP yield per O_2 consumed.[34] Thermodynamically, Cyd-II likely dissipates more free energy to increase flux rate and produces less proton motive force to drive ATP synthase and ATP–ADP disequilibrium.

Although Cyd-II is less efficient in ATP yield per food molecule and O_2 consumed, its greater respiratory flux may create greater efficiency of ATP production rate per unit membrane area. Comparatively,

glucose uptake → membrane competition → fast cytochrome.

Faster glucose uptake increases membrane demand for both glucose transporters and cytochromes, which favors greater expression of space-efficient Cyd-II relative to yield-efficient Cyo.

Cyd-I has greater affinity for oxygen than Cyo, raising Cyd-I's relative efficiency under low oxygen concentration.[34,413] Comparatively,

O_2 concentration ⊣ benefit of affinity → Cyd-I over Cyo.

Decreasing oxygen raises the benefit of affinity, which favors Cyd-I expression over Cyo expression.

At fast growth rates, if glucose transporters and cytochromes compete for limited membrane space, then knocking out the cytochromes should raise the glucose uptake rate,

cytochrome knockout ⊣ membrane competition ⊣ glucose uptake.

Similarly, when membrane area limits growth, raising the expression of space-efficient Cyd-II relative to the less efficient cytochromes may reduce membrane competition and increase glucose uptake,

Cyd-II expression ⊣ membrane competition ⊣ glucose uptake.

Shifts may also occur in the excretion rate of glycolytic products, typically acetate for *E. coli* under common lab conditions.

In eukaryotes, glycolysis in the cytosol leads to pyruvate uptake and electron transport on the mitochondrial membranes. Those compartments alter various surface and volume tradeoffs.[399,472]

Overall, the strongest gains and losses that shape traits may vary between limited time, limited food, limited proteome, or limited space.

COMBINING PROTEOME AND MEMBRANE LIMITATION

Basan et al.[28] claimed that proteome limitation dominates membrane limitation:

> Our findings in response to useless protein expression and energy dissipation are difficult to reconcile, even qualitatively, with ... constraints of the cytoplasmic membrane.[472]

Szenk et al.[399] do not claim victory for the membrane limitation theory over the proteome limitation theory, but they do argue for a way to a decisive win:

> A key prediction that distinguishes the membrane real estate hypothesis from ... [proteome limitation][28] is that it predicts that ... [the growth rate at which acetate overflow begins] is more sensitive to the overexpression of "dummy" proteins that reside in the inner membrane rather than the cytosol, as the former directly competes with the electron transport chain for membrane space.

The current data do not resolve these opposing theories. But is it sensible to present these two constraining limits as exclusive alternatives?

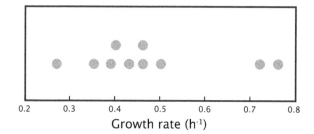

Figure 16.8   Growth rate at which acetate overflow begins in different experimental studies of *E. coli*. Varying lab conditions and different strains complicate the comparison between measurements. For our purposes, it is sufficient to note the significant variation between environments and genotypes. Redrawn from Fig. 4 of Szenk et al.[399]

This question poses a broader fundamental question: Should we be looking for the exclusive dominating constraint that shapes design?

The answer is often *no*. Different environmental challenges push organisms up against different limits. Alternative limits shift the dominant tradeoffs that shape design.

Proteomes and membranes may each limit respiratory flux under different conditions. The relative dominance of those constraining forces likely shifts with the environment. Some environments may push up against both constraints, shaping characters by the interaction between the different partial causes.

Consider, for example, the growth rate at which acetate overflow begins (Fig. 16.8). Both proteome and membrane limits may influence that switch point.

The proteome constraint depends on cellular volume, which limits the amount of cellular protein. The membrane constraint depends on the total cellular surface area.

As the volume of a regularly shaped cell increases, the surface to volume (S/V) ratio declines. That geometric relation influences the relative importance of the proteome and membrane constraints.

Faster *E. coli* growth associates with rising cell volume and decreasing S/V ratio (Fig. 16.9). That association between growth and cell geometry causes the proteome and membrane limitations to oppose each other,

S/V ratio → proteome limitation ⊣ overflow onset

S/V ratio ⊣ membrane limitation ⊣ overflow onset.

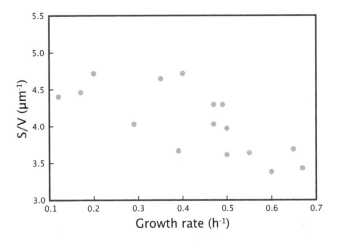

Figure 16.9   *E. coli* cell volume rises with growth rate, causing the surface to volume (S/V) ratio to decline. Redrawn from Fig. 1 of Szenk et al.[399] Original data from Volkmer & Heinemann.[430]

For the first prediction, rising volume and declining S/V ratio expand the proteome, reducing proteome limitation. Less proteome limitation weakens the tradeoff between growth and TCA flux (Fig. 16.3), allowing cells to increase growth rate to a higher level before facing the proteomic tradeoff that induces overflow.

For the second prediction, a declining S/V ratio reduces the membrane surface relative to the volume of catabolic flux. Greater membrane limitation causes overflow onset at a lower growth rate.

These two predictions highlight the opposing forces of proteome and membrane limitation. The interaction between those forces may influence the S/V ratio and overflow onset.

Electron transport can also influence proteome and membrane limitation and the onset of overflow metabolism. On the *E. coli* inner membrane, altering the balance of the various cytochromes changes the electron transport chain (ETC) rate.

The ETC rate measures the production rate of ATP per unit membrane area. Cytochromes with a faster rate have a lower yield of ATP per unit food input. The ETC rate may have opposing effects through the alternative constraints

ETC rate $\rightarrow$ proteome limitation $\dashv$ overflow onset

ETC rate $\dashv$ membrane limitation $\dashv$ overflow onset.

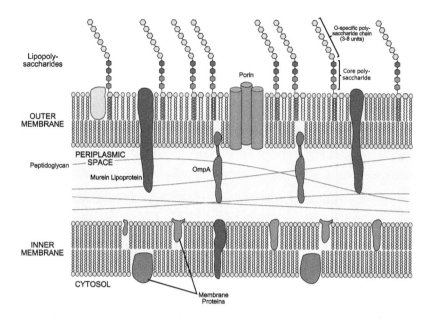

Figure 16.10   Outer and inner membranes in gram negative bacteria. OmpA and murein are abundant lipoproteins that provide structural integrity. From original by Jeff Dahl, Creative Commons CC BY-SA 4.0 license, via Wikimedia Commons.

Higher ETC throughput may increase TCA flux. More TCA flux raises catalytic demand, possibly increasing proteome limitation. Simultaneously, greater ETC rate reduces membrane limitation. These opposing partial causes may influence the growth rate at which overflow onset begins.

In this case, a rise in ETC rate associates with a decline in ETC yield. Different environmental conditions may favor rate or yield. When conditions favor different efficiency metrics, the balance of forces may change with respect to the constraining forces imposed by proteome and membrane limitations.

In summary, multiple constraints may act simultaneously when conditions push cells against physiological or biophysical limits. This subsection identified potentially important partial causes and possible interactions arising from proteome and membrane limitations.

OTHER LIMITS IMPOSED BY MEMBRANE SURFACE AREA

Membranes serve many functions. The biophysical links between those different functions create constraints and tradeoffs.[35,104,360,399] Additional constraints can influence overflow.[78]

Gram negative bacteria such as *E. coli* have an outer membrane and an inner membrane (Fig. 16.10). Porin molecules on the outer membrane create diffusion openings for external molecules (but see Ude et al.[416]). The inner membrane includes active transporters and the ETC system. I summarize and extend some tradeoffs mentioned by Szenk et al.[399]

- If outer porins limit nutrient uptake, then changes in surface area, porin density, and pore characteristics alter maximum nutrient flux.

- Reduced S/V in larger and faster growing cells lowers maximum nutrient uptake per unit volume. An increase in cell size to mitigate proteome limitation may increase the constraint imposed by the outer membrane diffusion barrier.

- Increased pore size may enhance diffusive nutrient uptake at the expense of greater uptake of toxins or other modes of attack.

- Reduced limits on diffusion imposed by the outer membrane increase the marginal gain for inner membrane transporters, raising the intensity of competition for space on the inner membrane between nutrient transporters and the ETC.

- Strong outer membrane diffusion limits reduce the benefit of additional inner membrane nutrient uptake, decreasing competition between nutrient receptors and the ETC.

- Porins and other diffusion pathways may be leaky, causing cells to expend much free energy to maintain concentration gradients across membranes.

- The more strongly free energy constrains cellular fitness, the more intensely a greater S/V ratio and leakier porins reduce cellular performance by dissipating free energy to maintain gradients.

16.5   Response to Environmental Challenge

TRADEOFFS CHANGE WITH CONDITIONS

With limited food, proteome metabolic demand remains below capacity. The extra capacity allows cells to prepare for new challenges.[51]

In *E. coli*, limited food and slow growth associate with greater proteome allocation to stress response. The primary stress response regulator RpoS increases in concentration with declining growth rate.[39] Stress tolerance rises.

Similarly, food-limited cells express proteins needed to catabolize many absent substrates.[191] Faster growth increases metabolic demand on the proteome, repressing pathways for absent food sources.

A three-way tradeoff arises for the proteome.[51] A catabolic part extracts free energy. An anabolic part builds new molecules. A look-ahead part prepares for new challenges.

This three-way partition illustrates how constraints and tradeoffs change with conditions. For example, when food is limiting,

$$\text{food} \rightarrow \text{metabolic gain} \dashv \text{look-ahead allocation.}$$

Increasing food enhances the gain from allocation to metabolic functions, which decreases the allocation to look-ahead preparation for stresses or alternative food sources.

With limited food, the proteome compartment does not constrain growth. As food continues to increase, metabolism and growth eventually become limited by the proteome. Catabolic and anabolic functions trade off against each other, potentially leading to overflow metabolism.

Proteome limitation can be increased experimentally by overexpressing a protein.[28] Increased proteome limitation predicts a lower growth rate at which the dominant tradeoff switches from metabolism versus look-ahead function to catabolism versus anabolism. The change in dominant tradeoffs with conditions is a primary theme of this book.

How much can forces of design alter the constraining tradeoffs? For example, more bouts of environmental stress or more frequent changes in the mix of available foods might enhance the benefit of the look-ahead allocation fraction,

$$\text{unpredictability} \rightarrow \text{look-ahead gain} \rightarrow \text{growth vs look-ahead.}$$

Greater environmental unpredictability increases the marginal gain for look-ahead function, which strengthens the proteome tradeoff between allocation to growth versus allocation to look-ahead functions.

METABOLISM VERSUS OXIDATIVE STRESS RESISTANCE

In fission yeast, a three-way tradeoff arises between growth rate, biomass yield, and oxidative stress resistance (p. 172).[202]

Genotypes with a relatively more active pyruvate kinase isoform increase glycolytic flux. Cells grow faster, produce lower biomass yield, and overflow more glycolytic fermentation products.

The fast genotype also reduces oxidative stress tolerance, measured by exposure to hydrogen peroxide or diamide.

Among 161 natural isolates, 143 have the fast pyruvate kinase variant. The commonly studied lab strain is among the 18 with the slow variant.

In the lab strain, replacing the slow variant by the fast variant altered phenotype toward fast growth, low yield, and reduced stress tolerance. Thus, pyruvate kinase activity explains at least part of the phenotypic differences between genotypes.

Why does nature maintain genetic variation in pyruvate kinase? Do varying forces alter the relative benefit of rate versus yield, as discussed in earlier chapters?

Does varying oxidative stress tolerance arise indirectly as a correlated response to forces acting on metabolism? Or do naturally varying forces act directly on oxidative stress tolerance?

The current data give few clues. Making a list of comparative predictions provides a start. Many predictions will be wrong. Experts will think of better predictions. Nonetheless, predictions are the way forward.

The following subsections focus on proteome limitation, pathway flux control, the primary redox disequilibria that drive cellular physiology, and the relation between redox reducing power and oxidative stress resistance. These factors provide insight into metabolic design.

PROTEOME LIMITATION

Proteome limitation might connect metabolism to stress resistance, for example,

$$\text{rate} \rightarrow \text{proteome limit} \dashv \text{stress tolerance}.$$

Faster growth rate increases proteomic demand, which limits allocation to oxidative stress tolerance. Manipulating the proteome size may reveal how a constrained proteome connects growth rate to stress resistance.

Some facts about pathway flux and oxidative stress resistance suggest more specific predictions.

PPP FLUX AND THE NADPH DISEQUILIBRIUM

The pentose phosphate pathway (PPP) yields precursors for biosynthesis and plays an important role in other metabolic functions (p. 180).[394] Here, I focus on the PPP's role in oxidative stress resistance. I emphasize results from commonly studied yeast species fed on glucose.[163,202]

Incoming glucose is transformed into glucose 6-phosphate, which may continue through the PPP or through alternative glycolytic pathways. The initial steps of the PPP produce two NADPH, increasing the NADPH–NADP$^+$ disequilibrium (Fig. 16.11).

The NADPH–NADP$^+$ disequilibrium provides the redox reducing power that ultimately drives electrons toward oxidatively damaging molecules, such as hydrogen peroxide, neutralizing oxidative stress as in the transformation

$$H_2O_2 + 2\,e^- + 2\,H^+ \longrightarrow 2\,H_2O.$$

The actual cascade may begin with NADPH passing electrons to glutathione, which acts directly as the antioxidant that reduces hydrogen peroxide to water.

The NADPH–NADP$^+$ disequilibrium also drives many important biosynthetic pathways, such as the production of various lipids, nucleic acids, and sugars. When the PPP is knocked out, cells can typically maintain a sufficient NADPH–NADP$^+$ disequilibrium to drive biosynthesis.

PPP negative cells fail to tolerate oxidative stress, suggesting that cells need the NADPH created via the PPP for antioxidant function. The PPP may also upregulate other essential antioxidant and repair mechanisms, such as providing nucleotides for repairing oxidative DNA damage.

RESPONSE TO OXIDATIVE CHALLENGE

In budding yeast, an oxidative challenge causes a rapid shift in flux toward the PPP.[163] On the order of seconds, oxidation of the enzyme glyceraldehyde 3-phosphate dehydrogenase inhibits flux in upper glycol-

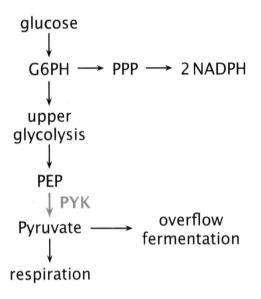

Figure 16.11   Alternative flux pathways through glycolysis and the PPP. Greater PPP flux enhances oxidative stress resistance. A slower variant of pyruvate kinase (PYK) reduces glycolytic flux and enhances PPP flux. Abbreviations: glucose 6-phosphate (G6PH), pentose phosphate pathway (PPP), phosphoenolpyruvate (PEP).

ysis, shifting flux to the PPP. On the order of minutes, the rapid initial increase in PPP flux is supported by enhanced gene expression of glucose 6-phosphate dehydrogenase (G6PDH), the first catalytic step from G6PH into the upper branch of the PPP (Fig. 16.11).

PYRUVATE KINASE VARIANTS ALTER PPP FLUX

The final step in glycolysis transforms phosphoenolpyruvate (PEP) into pyruvate, catalyzed by pyruvate kinase (PYK). As noted above, high activity and low activity variants of PYK occur. Budding yeast carry both high and low activity PYK isoforms, which can be alternatively expressed.

When comparing high activity versus low activity PYK, the greater catalytic activity increases flux through the later glycolytic steps. That greater flux through pyruvate associates with reduced PPP flux, higher sensitivity to oxidative stress, more overflow of fermentation products, and reduced aerobic respiration.

The low activity variant causes greater PPP flux and improved oxidative stress tolerance. Low activity PYK also associates with greater respiration and higher flux through oxidative phosphorylation, which may increase intrinsic oxidative stress.

Most eukaryotes[394] and some bacteria, including *E. coli*,[469] have multiple genes for pyruvate kinase. The different isoforms can be variably expressed. Those isoforms may allow physiological tuning of fermentation, respiration, and oxidative stress resistance.

Tuning of Flux and Oxidative Stress Resistance

Interestingly, fission yeast have only a single *pyk* gene.[290] High and low activity variants occur polymorphically among natural isolates (p. 283).

Most individual microbial cells physiologically tune metabolic flux and oxidative stress resistance by changing expression of PYK isoforms. By contrast, fission yeast populations evolutionarily tune their response by changes in the frequency of alternative *pyk* alleles.

Comparative Predictions for Species with PYK Isoforms

We know relatively little about the dominant natural causes of oxidative stress.[192] Certain physical processes create reactive oxygen species (ROS) in the environment. Competitors often deploy ROS. Higher oxygen concentration increases endogenous cellular generation of ROS.

Whatever the cause, comparatively,

$$\text{oxidative stress} \rightarrow \text{NADPH benefit} \rightarrow \text{PPP flux.}$$

Greater oxidative stress enhances the benefit of the NADPH–NADP$^+$ disequilibrium, which favors increased PPP flux.

The correlations of PPP flux with growth rate, yield, respiration, and overflow suggest related predictions. For example, increased competition between different genotypes may favor faster growth rate at the expense of reduced yield. Mechanistically, faster growth often arises from greater glycolytic flux and fermentation overflow, which may require faster PYK catalysis. Comparatively,

$$\text{relatedness} \dashv \text{competitive gain} \rightarrow \text{fast PYK.}$$

More genetic mixing and lower relatedness enhance the competitive gain of growing faster than neighbors, which favors more rapid glycolytic flux and faster PYK activity. In the same way, various demographic changes that increase the gain for fast growth also favor fast PYK.

Fast PYK correlates with increased overflow, reduced respiration, lower PPP flux, and lower tolerance of oxidative stress.

These multiple correlated traits lead to alternative partial pathways of causation. For example, the causal pathways may begin by changes in oxidative stress or by changes in genetic mixing between competitors.

Other partial causes may occur. Suppose, for example, that an over-flow fermentation product such as acetate increased significantly in environmental concentration. Cells may no longer be able to drive excretion of the fermentation product. Or there may be a reverse flow back into the cell.[96,278] Comparatively,

overflow concentration ⊣ overflow growth benefit → growth rate.

High external concentration of overflow products reduces the benefit of overflow for growth rate, which lowers the potential growth rate. Consequently, the benefit of the fast PYK isoform for increased growth rate declines, potentially favoring the slower isoform, slower growth, more respiration, and greater oxidative stress tolerance.

EVOLUTIONARY TUNING OF PHYSIOLOGICAL RESPONSE

A burst of oxidative challenge reduces the catalytic activity of enzymes in upper glycolysis. That reduced upper-glycolytic activity shifts flux to the PPP (Fig. 16.11), causing a fast rise in the NADPH–NADP$^+$ disequilibrium and the detoxification of ROS.

Some cells initially express a fast PYK isoform and reduced oxidative stress tolerance. An oxidative challenge may trigger those cells to switch to a slower PYK isoform that enhances PPP flux and ROS detoxification.

Switching isoform expression associates with various changes in gene expression and flux. Those broad cellular responses happen relatively slowly (minutes). By contrast, rerouting flux through PPP in response to oxidative block of upper-glycolytic catalysis happens relatively quickly (seconds).[163]

How might design forces evolutionarily tune the physiological response of switching gene expression between PYK isoforms? In other words, what shapes the phenotypically plastic response?[82,139,321,356]

Consider oxidative challenge as a time varying input signal. If challenge happens as short bursts, with little temporal correlation between a burst and a subsequent challenge, then a switch from a fast to a slow isoform should not begin in response to a short oxidative burst.

By contrast, if an initial burst commonly associates with a longer challenge period, then a rapid start to isoform switching may be favored in response to an initial burst. In general, the temporal characteristics of the challenge input signal may evolutionarily tune the response pattern. Comparatively,

informative signal → response benefit → isoform switch.

The more an input signal informs about a changed environment, the greater the response benefit and associated switch in isoform expression.

Control theory provides a framework to analyze signal information and response benefits (Chapter 7). Predictions may be tested by comparing environments with different signal information properties. Or experimental evolution may study how altered signal information changes the physiological response.

COMPARATIVE PREDICTIONS FOR SPECIES WITH *pyk* GENETIC VARIANTS

Fission yeast have only a single *pyk* gene, as noted above. Fast and slow allelic variants occur among natural isolates.

Environments that favor oxidative stress tolerance or high yield select the slow allele over the fast allele. Environments that favor rapid growth and impose limited oxidative challenge select the fast allele.

Environments inevitably vary. The theory of fitness in variable environments provides predictions (Section 5.10).

If environmental fluctuations affect all individuals carrying a particular allele in the same way, then the average fitness of an allele is significantly discounted by its variability in fitness, favoring the allele with the highest geometric mean fitness.

By contrast, if individuals carrying a particular allele experience or respond to environmental fluctuations in an uncorrelated way, then the allele with the highest average fitness tends to dominate. Comparatively,

correlation → benefit of reduced variance → geometric dominance.

Greater fitness correlation between individuals with the same allele raises the benefit of reduced variance in fitness, causing greater dominance by the allele with the higher geometric mean fitness.

Equivalently, greater independence of fitness between individuals with the same allele lowers the benefit of reduced fitness variance and raises dominance by the allele with the highest arithmetic mean fitness.

Outcome depends on demography. In this example, individuals respond to environmental fluctuations, then compete globally for transmission to the next generation. Alternatively, local competition would alter how variability in fitness and correlations between individuals affect relative allelic success.[131,144]

INDIVIDUAL RESPONSIVENESS VERSUS GENETIC VARIATION

What causes some populations to respond by individual adjustments to the environment and other populations to respond by changes in allele frequencies? Matching each individual's response to its environment would seem to dominate, all else being equal. What might not be equal?[82,321,356]

Perhaps the proteomic demand or other costs associated with sensing the environment and switching expression outweigh any benefits. Or switching isoform expression and the associated pathways may happen more slowly than environmental fluctuations.

In budding yeast, which has alternative isoforms, knockout strains could be created that have only one of the isoforms.

Experiments could compete the multi-isoform and single-isoform types under different conditions. Although the comparison is artificial, it may be possible to find environmental conditions and evolutionary changes that favor single-isoform types. Follow-up may yield clues.

DESIGN

This section emphasized two constraints. Proteome limitation imposes a tradeoff between stress response proteins and catabolic enzymes. Glycolytic flux limitation imposes a tradeoff between PPP flux and alternative pathway flux. PPP flux increases oxidative protection by enhancing the $NADPH-NADP^+$ disequilibrium at the expense of slower catabolic flux.

Those constraints impose physiological limits on metabolism. The relative benefit of the growth rate and biomass yield components of fitness may modulate metabolic design significantly within those constraints.

For example, mixing of genotypes enhances the benefit of fast growth to outcompete neighbors. That enhanced benefit may favor an evolutionary shift in the relative proteome allocation from stress resistance to catabolic flux. Greater benefit for fast growth may also shift the balance from PPP flux to alternative glycolytic flux.

16.6 Summary

The fundamental duality between the forces of constraint and design recurs in every case study. Physiological and biophysical constraints set boundaries on the possible. Within those boundaries, natural selection often has wide scope for modulating design.

The constraining boundaries may be modified. For example, changes in cell size alter proteome limitation. Changes in membrane porosity alter the way surface area constrains respiration.

17 Predictions: Diauxie, Electrons, Storage

This chapter links Part 1's theory for biological design to Part 2's problems of microbial metabolism.

The first section analyzes diauxie, the switch between alternative food sources. Microbes typically prefer some foods over others. The more strongly a cell focuses on a preferred food and represses the catabolic machinery for an alternative food, the longer it may take to switch to the second food after the first has been consumed.

How does temporal and spatial variation in food availability shape catabolic regulation? That question links metabolic design to theories of demography, life history, kin selection, and evolution in variable environments.

The second section considers the catabolic need for a final electron acceptor. Catabolic free energy comes from moving the weakly held electrons in food to strong electron acceptors, such as oxygen.

Finding an electron terminus becomes a primary catabolic challenge in the absence of oxygen or another abundant electron acceptor. Some microbes solve this challenge by transporting catabolic electrons to distant acceptors outside the cell.

Distributed electron flux raises many interesting problems of design. For example, some cells may produce extracellular electron shuttle molecules that carry electrons from the cell surface to a distant acceptor.

The extracellular shuttles, once released, may be used by neighboring cells. Such publicly shareable resources set many interesting design challenges. This section develops aspects of spatial scale and competition that arise for shareable public goods.

The third section evaluates cellular storage in complex life cycles. Cell lineages often pass through multiple habitats. Some habitats may lack food or final electron acceptors, preventing catabolism and requiring the use of stored free energy.

Building up and using resource stores often imposes tradeoffs. In a habitat that prevents catabolism, a bit more stored free energy used for growth means less available to maintain survival if the famine continues. The life cycle influences the relative costs and benefits for growth versus survival and thus for building and using storage.

This third section develops a wastewater treatment example that links alternating habitats and metabolic biochemistry to theoretical concepts of microbial design. Industrial microbiology provides many excellent models to study the forces that shape design.

The final section relates the particular models of microbial life history in this chapter to the broader problems of biological design. The primary challenge often concerns how to match the changes that we can study directly by observation or by dynamical models to the underlying forces that explain the motion, leading to Lanczos' advice "to focus on the forces, not on the moving body."[222]

17.1 Switching between Food Sources

Microbes often prefer particular foods (Section 14.5). For example, *S. cerevisiae* feeds first on glucose. It then shifts to feeding on other food sources, such as galactose.[98,340]

While feeding on a preferred sugar, cells that express catabolic genes for an alternative sugar may suffer reduced growth.[437] In spite of that cost, pathway expression for the inferior food occurs in some cells.[340]

What forces shape mixed expression? This section illustrates how the core theory of Part 1 leads to comparative predictions.

PATCH LIFESPAN AND CYCLE FITNESS

Suppose a habitat divides into several isolated resource patches. Each newly formed patch contains an initial allocation of two sugars. No new sugar flows into an existing patch.

The first sugar provides a better food source for growth than the second one. How do design forces shape the pattern by which microbes consume the two food sources?

Comparatively, shorter patch lifespan favors high growth rate over efficient yield (Fig. 4.1). Intuitively, if a patch disappears before the resources are used up, efficient yield provides little benefit.

This age-specific force shapes how cells use the two different sugars. In this case, *age* means the time passed since a patch was colonized.

Age-specific forces can often be analyzed by demographic cycle fitness (Section 5.7).

A simple model links catabolic regulation to cycle fitness. Existing patches die in each small time increment at a rate λ. New patches arise at the same rate. Each patch contains genetically homogeneous microbes, with genetic variation between patches.

In each time increment, migrants from existing patches colonize the new patches. All migrants to a new patch come from a single existing patch. The probability that an existing patch produces successful migrants is proportional to the current patch biomass, which measures population size and migration potential.

Demographic cycle fitness for a genotype is the sum, over all time increments of its patch's lifespan, of the patch biomass at a particular time, $b(t)$, multiplied by the probability of surviving to that time, t, as

$$w = \int b(t)\lambda e^{-\lambda t}dt,$$

with average patch lifespan, $1/\lambda$. The caption of Fig. 17.1 describes the dynamics for the availability of the two sugars, the proteome expression levels for the pathways to catabolize each sugar, and the biomass.

The dynamics follows an intrinsic tradeoff between growth rate and biomass yield. Mechanistically, higher catabolic expression allocates more resources to increase growth rate, reducing resources available to produce biomass yield.

For each of four average patch lifespans, $1/\lambda = 1, 2, 4, 8$, I maximized fitness subject to the rate-yield tradeoff mediated by catabolic expression level. In Fig. 17.1, the increasingly lighter shading of the curves corresponds to increasingly longer patch lifespans.

As patch lifespan increases, growth rate slows and biomass yield increases (Fig. 17.1a). In addition, the lag between consumption of the two sugars declines.

The catabolic expression levels explain the reduced lag (Fig. 17.1b). The top and bottom sets of curves show expression levels for the first and second sugar, respectively.

With longer patch lifespans, expression level for the primary sugar increases more slowly. Lower expression reduces growth rate, lowers the proteome cost for catabolism, and raises the biomass yield.

Proteome expression for the first sugar represses catabolic expression for the second sugar. As the expression level for the first sugar declines

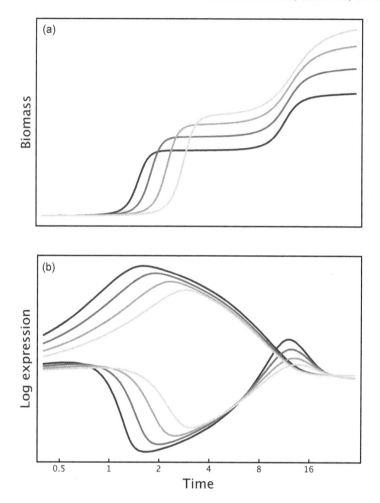

Figure 17.1   Patch lifespan alters the sequential consumption of two sugars. Average patch lifespan increases with lighter shading of curves. (a) Biomass change with time shows the growth rate during different periods of sugar consumption. (b) Expression level shows the proteomic allocation to the catabolic cascade. The top set of curves in this panel corresponds to expression for catabolism of the first sugar, the bottom set for the second sugar. See text for explanation of dynamics, which is given by the variables, x_i, y_i, b, for sugar concentration, proteome expression, and biomass, with $i = 1, 2$ for the first and second sugar, and tildes for initial values, with $\dot{x}_i = -x_i y_i b$, and $\dot{y}_i = y_i x_i y_i (f - y_i/5) + (\tilde{y}_i - y_i)/2 - c_i y_1 y_2$, and $\dot{b} = b(x_1 y_1 + 0.9 x_2 y_2)(1 - f)$, with $y_1 = 1, y_2 = 0.35, c_1 = 0, c_2 = 12, r_1 = 1, r_2 = 0.9, \tilde{x}_i = 10, \tilde{y}_i = 0.01, \tilde{b} = 0.1$. The value of f maximizes fitness, calculated by $\int_0^{30} b(t) \lambda e^{-\lambda t} dt$ for average patch lifespans $1/\lambda = 1, 2, 4, 8$. This model extends Frank's[130] analysis for a single resource.

with longer patch lifespan, the catabolic expression for the second sugar rises.

During the initial feeding on the first sugar, greater expression for the second sugar reduces the lag between growth on the two sugars.

Empirically, two key puzzles of sequential sugar usage often arise. Why is significant expression level for the second sugar sometimes maintained during consumption of the first sugar? What explains variability in the lag time between growth on the first and second sugars?

Testable comparative predictions follow from the simple patch lifespan model. For example, increasing patch lifespan emphasizes yield over rate, which favors a more balanced catabolic expression for the two sugars and shorter lags between growth on alternative food sources.

COMPETITION AND RELATEDNESS

Figure 17.1 assumes a single type in each patch. If the types mixed and competed within patches, that competition would favor faster growth and lower yield (Section 4.1).

The relatedness between competitors measures the intensity of competition (Section 5.2). The more mixing of types during patch colonization, the lower the relatedness and the more intense the competition between different types for local resources.

How does increasing competition between types affect the consumption of multiple resources? Chapter 5 provides the concepts to analyze relatedness and demography. I limit comments here to likely comparative tendencies based on a simple qualitative approach.[130]

Lower relatedness and greater competition favor faster growth rate and reduced biomass yield. In Fig. 17.1, decreasing relatedness has a similar effect to shorter patch lifespan (Fig. 4.1).

The benefit of faster growth alters the regulation of sequential resource consumption. The particular change depends on the tradeoffs imposed by the mechanistic basis of regulation.

The commonly studied lab species are often tuned for competition and fast growth. By contrast, natural environments vary widely in demography and relatedness. Broad comparative changes likely occur.

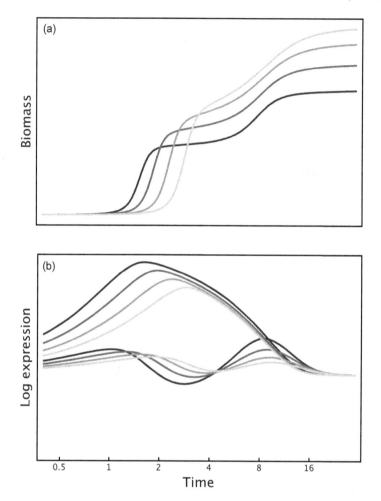

Figure 17.2   Reduced proteome limitation allows higher catabolic expression for a second food source, shortening the lag time for the switch between food sources. Same model as in Fig. 17.1, with $c_2 = 2$ to reduce the repressive effect of catabolic expression for the first food source on the expression level for the second food source.

PROTEOME LIMITATION

In Fig. 17.1b, catabolic expression for the first food strongly represses expression for the second food. Repression may arise because of proteome limitation, which imposes a tradeoff between catabolic protein expression for the alternative foods.

What happens under reduced proteome demand, which weakens the tradeoff between catabolic expression for alternative foods?

To study that question, Fig. 17.2 analyzes the same model, with a reduced intensity for the repressive effect of the catabolic proteins for the first food on the catabolic expression for the second food.

Figure 17.2b shows that less intense proteome limitation allows maintenance of greater catabolic expression for the second food source while consuming the first source. That greater expression reduces the lag time between growth on the alternative food sources (Fig. 17.2a).

As always, this comparative prediction describes a partial causal effect. Reduced proteome demand may also affect other causes. Those other causes may alter the net effect on catabolic expression.

If so, then reduced proteome demand may sometimes fail to increase the catabolic expression for the second food source. However, the tendency over different cases should be in the predicted direction.

UNPREDICTABLE RESOURCE INFLUX

The prior examples assumed that each resource patch starts with a fixed amount of the two alternative sugars. What if the temporal and spatial patterns of resource flux vary? Six simplified scenarios highlight important partial causes (Section 5.10).

Environmental fluctuations favor catabolic expression that buffers against variability in performance.—Suppose the environment divides into many separate patches. Each patch contains a single genotype. At the start of each of n identical time intervals, a random amount of each sugar arrives and is split equally among the patches.

After the final time interval, all patches contribute migrants to a global pool. Each patch contributes migrants in proportion to its final population size.

The current patches disappear. New patches arise. One genotype colonizes each new patch. A genotype's frequency in the migrant pool determines its colonization success.

In this scenario, what pattern of catabolic expression improves success? To answer, we must consider the sequence of growth in each patch in response to the random influx of additional sugar.

Suppose population size in a patch increases by a multiplicative factor λ_{ij} in time interval i for genotype j. Total reproductive yield for genotype

j over a complete demographic cycle is the product of the λ_{ij} values over the sequence of time intervals (Section 5.10).

A multiplicative product of values scales with the geometric mean of those values. Thus, the total-yield measure of fitness over a complete demographic cycle scales with the geometric mean of growth values over the sequence of time intervals.

In terms of regulatory control, the forces of design favor catabolic expression for the two sugars that increases geometric mean growth. The geometric mean rises with the arithmetic average of the interval growth values, λ_{ij}, and declines with variability in those values.

Comparatively, more variable mixtures of sugars often favor more balanced catabolic expression. Balanced catabolic expression enhances the geometric mean growth when the gain for reducing the variability in growth outweighs the loss for reducing the arithmetic average.

For example, more variable resource influx will sometimes present cells with a relatively high amount of the less preferred sugar. To prevent low efficiency because of unused high expression for the preferred sugar, cells may tend to balance their initial expression for the two sugars.[380]

That bet-hedging[386] between alternatives relates to timescale. The highest geometric mean maximizes the expected growth rate over the full demographic cycle. A higher expected success over the full cycle may associate with lower success in particular intervals of resource influx.

Spatially correlated variability and global competition.—In the prior example, the final population size of each patch is the product of growth during each episode of resource influx. That multiplication of periodic growth leads naturally to a geometric mean measure of fitness. After several episodes of local resource influx and growth, migrants mix and recolonize new patches.

An alternative demographic cycle may occur. New patches form, each containing the same mixture of the two sugars. The particular mixture varies at the start of each cycle. Microbes grow by consuming the initial sugar allocation. Following a period of growth, the microbes disperse, mix globally, and colonize new patches.

Suppose there are two genotypes. A focal genotype has frequency q. The alternative has frequency $1 - q$. After one cycle, the frequency of the focal genotype is

$$q' = \frac{w_1 q}{w_1 q + w_2 (1 - q)}, \tag{17.1}$$

in which w_1 is the multiplicative fitness factor by which the focal geno-type increases its population size during its period of growth within its patch, and w_2 is the fitness for the alternative genotype.

Because the resource mixtures vary over time, the fitness values will also vary. As in the prior scenario, we assume that the different geno-types have different patterns of catabolic regulatory expression for the alternative sugars. Those different expression patterns cause differences in fitness between the genotypes.

Once again, the favored genotype has the highest geometric mean for its temporal sequence of fitness values. For example, suppose the resource influx has two equally probable states with fitnesses $w_1 = 1 \pm \delta$ and $w_2 = 1 \pm (\delta + \epsilon)$, with $\delta, \epsilon > 0$.

The arithmetic average fitness for both genotypes is 1. The geometric mean fitness is greater for w_1 because of its lower variability in fitness values. The genotype with the higher geometric mean fitness increases against the alternative genotype.[80,144,153]

A genotype with a lower arithmetic average fitness can increase if its fitness varies less between environments, leading to a higher geometric mean. If, between two environments, $w_1 = 1$ for both environments and $w_2 = \{0.82, 1.2\}$ for the two environments, then the arithmetic and geometric means for w_1 are both 1, and for w_2 are 1.01 and 0.992.

In this case, the first genotype typically increases in frequency because it has the higher geometric mean fitness, in spite of its lower arithmetic mean fitness. Using eqn 17.1 with an initial frequency of $q = 0.5$, after one episode of each environment, $q = 0.504$.

Dominance of the geometric mean implies that genotypes may gain by trading a reduced arithmetic mean success for a reduced variability in success across environments.

Comparatively, the more variable the environment over time, the more strongly catabolic regulation may trade reduced arithmetic success for reduced variability in success.

Spatially uncorrelated variability and global competition.—In the prior examples, resources flow into each patch in the same way. Alternatively, resource flows may be uncorrelated between patches.

If much of the variability in resource flows occurs spatially, then a genotype experiences nearly all of the alternative environments in each time period. The total contribution of that genotype to the migrant pool at the end of the growth cycle depends on the arithmetic average success over the different spatial environments.

If arithmetic average success over spatial patches varies little over time, then the geometric mean success of the genotype over time is very close to the arithmetic average over space.

Comparatively, the more weakly correlated resource flows are between patches and the more closely the spatial distribution matches the temporal distribution of patch characteristics, the more strongly the arithmetic mean will dominate over the geometric mean.

The particular catabolic expression patterns that maximize arithmetic versus geometric mean success depend on the details. Typically, the geometric mean favors reduced variability in success.

Thus, less spatial correlation implies greater dominance by the arithmetic mean, less emphasis on reducing variability in success, and less bet-hedging in the expression of catabolic regulation.

Stochastic gene expression reduces spatial correlation.—Catabolic expression may vary stochastically between individuals of the same genotype.[380] Such variability can reduce the spatial correlation in fitness.

Lower correlation between individuals tends to reduce the variability in genotype success. Lower variability raises the geometric mean.

Comparatively, increased environmental variability favors greater stochasticity in gene expression to reduce genotypic variability and raise geometric mean fitness.

That comparative prediction highlights a partial causal effect. The particular details influence the overall value of stochastic variability.

Local competition for colonization and rare-type advantage.—Prior examples emphasized spatial correlations in environments and in trait expression between individuals. Spatial properties define demographic aspects of populations.

This example focuses on the spatial scale of competition, another key demographic property. As before, suppose that each patch begins with a single genotype and a random allotment of the two sugars. Cells consume the sugars and increase their population. A migrant pool forms, colonizes new patches, and repeats the cycle.

In prior examples, all new patches receive their colonists from a global migrant pool. In this example, the global set of patches divides into many distinct regions.

Each region forms its own migrant pool, which then colonizes only the newly formed patches in the local region. Occasionally, a global cycle of migration and colonization occurs. Most of the time, competition for colonization happens locally within regions.

Comparatively, greater local competition increases the amount of genetic variability maintained in the global population.

That increase in genetic variability occurs because environmental fluctuations favor rare types. Equation 5.19, repeated here, illustrates the rare-type advantage,

$$\mu_1 - q_1 \rho_1 \sigma_1^2 > \mu_2 - q_2 \rho_2 \sigma_2^2.$$

The two competing genotypes, with subscripts $i = 1, 2$, have arithmetic fitness means, μ_i, frequencies, q_i, fitness correlations between individuals of the same genotype, ρ_i, and variances in fitness, σ_i^2.

This expression describes the conditions for genotype 1 to gain against genotype 2, assuming that fitness fluctuations are small and the correlation between genotypes is zero.

The smaller a genotype's frequency, q_i, the less the overall fitness discount that arises from the variability in success, σ_i^2. That rare-type advantage tends to maintain genetic variability. For example, if we assume equal arithmetic means and equal correlations, then overall fitnesses in the above expression become equal when

$$\frac{q_1}{q_2} = \frac{\sigma_2^2}{\sigma_1^2}.$$

Less variability in success increases the equilibrium frequency of a genotype.

However, when competition occurs in a single global population, fluctuations in frequencies often cause extinction of the genotype with more variable success. Ultimately, the less variable genotype tends to dominate the global population.

By contrast, when competition for colonization happens locally across many separate regions, the global frequencies arise by averaging over the local regions. Averaging many independent local events lowers the global fluctuations. With lower fluctuations in frequencies, the equilibrating tendency of the rare-type advantage dominates, maintaining genetic diversity.

Local competition for resources.—The prior examples assumed a single genotype in each isolated resource patch. Mixing genotypes in patches would impose local competition for resources. Such local competition typically favors higher growth rate at the expense of reduced yield.

With regard to variable environments, within-patch competition alters how patterns of catabolic expression affect arithmetic mean fitness and the fluctuations in fitness. Mixtures within patches also induce correlations between genotypes with regard to environmentally caused fluctuations in fitness.

It would be interesting to develop theory that combined direct competition between genotypes and spatially induced correlations in fitness. Both causal paths influence the favored patterns of catabolic expression.

17.2   Distributed Electron Flux

Catabolism generates free energy by passing electrons through a redox gradient. Roughly speaking, the initial food source holds electrons relatively weakly. The final electron acceptor holds electrons relatively strongly. The electron gradient flows from low to high entropy or, equivalently, from high to low free energy.[467]

The catabolic free energy potential depends on the strength and availability of the final electron acceptor. Finding a strong, renewable receptacle for electrons forms a primary challenge of metabolism.

Many organisms pass electrons to oxygen, the basis for aerobic respiration. Some habitats lack sufficient free oxygen. Alternative electron acceptors create broad metabolic diversity in microbes.

Electron flux typically happens between coupled biochemical reactions within cells. However, some microbes move free electrons from donor molecules to spatially distant receptor molecules, creating an extracellular electric current (Section 14.4).

CABLE BACTERIA

Cable bacteria transmit electric currents across thousands of connected cells.[277] The current passes electrons to oxygen or nitrate. The terminal cells act solely as conduits, failing to grow.[151]

Cooperation between the connected cells and the reproductive sacrifice of the terminal cells pose interesting puzzles (p. 225). Before turning to those puzzles, I briefly review the biology (p. 213).

Genomic analysis suggests a broad array of catabolic pathways in cable bacteria.[210,289] A particularly interesting catabolic cascade follows two steps. First, in a habitat without free oxygen, cells feed on hydrogen sulfide by splitting that molecule with water to make sulfate, free protons, and free electrons (eqn 14.2).

Second, multiple cells physically link to form a cable that stretches from the anoxic zone to a region that has free oxygen.[151,152,210,368] Electrons flow along nickel-protein wires in the cellular periplasmic space between the inner and outer membranes, creating an electric link from the anoxic zone to the oxic zone.[45] At the oxic terminal, the incoming electrons combine with oxygen and protons to make water (eqn 14.3).

More than 90% of the cells reside in the anoxic zone. The remaining nonreproductive cells in the oxic zone act solely as wires that conduct electrons to the oxic region. Why would the oxic cells act altruistically, providing a benefit to the anoxic cells without gaining any growth benefit for themselves?[151]

The unusual natural history raises other questions. How do the cables form and grow? Do long cables sometimes split into smaller cables? If so, how do the buds compete with each other spatially? How do the bacteria colonize new resource patches?

Limited empirical information prevents convincing comparative predictions. However, more observations will soon follow because cable bacteria occur widely in aquatic sediment and may play a key role in geochemical cycles.[53,248,342,361]

I first summarize the details of metabolism and the life cycle. From that sketch of the natural history, I then consider the forces that may shape metabolic design. Compelling puzzles of microbial life history arise.

Altruistic oxic terminal cells.—Cells in the oxic zone do not assimilate carbon and do not divide.[151,152] The rapid electron influx typically reduces oxygen to water, perhaps causing strong oxidative stress.[210,368]

Oxic zone cells are likely related to their connected cable partners by recent cell divisions, causing high genetic similarity between cells in a cable. That high genetic similarity provides the most likely explanation for the oxic cells' sacrifice of their own reproduction to benefit their anoxic partners (Section 5.2).

Other processes may explain oxic cellular sacrifice. Alternatives include randomization with regard to their role as contributor to or recipient of benefits to neighbors, synergistic fitness benefits between partners with low relatedness, and positive ecological feedback.[126] However, initial focus should be on the simple explanation of high relatedness before considering more complex processes.

The forces that shape cable bacteria altruism and life history depend on how cables form, grow, compete, and disperse. A few limited studies provide initial observations.

Cable growth, cell number, and length.—Geerlings et al.[152] observed cable growth in the lab. Cables grew rapidly when connected to both the anoxic electron source and the oxic electron sink. Loss of contact at one terminal stopped growth.

Cell length is approximately 2–3 μm.[259,354] Cable length increases up to a few cm, with about 4000 cells cm^{-1}.

Movement.—Bjerg et al.[41] observed cable movement in slide preparations. The center of the slide contained anoxic sediment. Slide edges provided an oxic interface. Cables actively moved toward the oxygen boundary. The leading filament tended to stop when in contact with the oxic zone. As the oxygen boundary moved, cables followed.

Cables may elongate from an oxic zone boundary down into the anoxic sediment. Cables are relatively straight near the oxic boundary. Cables may coil into a terminal snorkel toward the sulfide electron source.[354,462]

Population density.—Schauer et al.[354] collected deep sediment, likely below the zone at which cables could connect to the oxic zone. They incubated the sediment with the upper layer exposed to oxygen. The initial cable bacteria density was below their threshold detection for cables at 10^4 filaments cm^{-3} or, for single cells, at 1.5×10^6 cells cm^{-3}.

At initial incubation, anoxic-zone sulfide occurred within 1 mm of the anoxic-oxic interface. After 10 days of incubation and consumption by the growing cable bacteria, most sulfide retreated to a boundary 2–8 mm below the oxic zone, dropping to 20 mm by day 53.

In the 0–15 mm depth interval, at 10 days, the cable bacteria density in each cm^2 increased to approximately 1 km of cables. At 21 days, density per cm^2 approximately doubled to over 2 km, or about 8×10^8 cells cm^{-3}.

Habitat heterogeneity and fragment length distribution.—Various physical and biotic processes perturb sediments. Disruption may break bacterial cables, preventing growth.[259] Disruption may also create new opportunities to link anoxic sulfide zones to oxic zones.

For example, parchment worms build tubes into the sediment, creating novel oxic zones. Abundant cable bacteria occur near those tubes. The tubes can be structurally stable for several months.[8]

Measured cable lengths near tubes were up to 1 mm, approximately 400 cells. A few mm away from tubes, measured cable lengths tended to be 5-15 cells. However, this study's measurement methods could not rule out fragmentation during sample processing.[8]

Overall, the size distribution of unconnected cables remains an open problem. A recent lab culture found single cells and short cables in a growing population[289] (see below).

ATP generation.—Overviews of cable bacteria metabolism typically emphasize the half reactions in eqns 14.2 and 14.3, repeated here,

$$H_2S + 4H_2O \longrightarrow SO_4^{2-} + 8e^- + 10H^+ \qquad (17.2a)$$

$$2O_2 + 8H^+ + 8e^- \longrightarrow 4H_2O. \qquad (17.2b)$$

The first reaction, in the anoxic zone, hydrolyzes hydrogen sulfide. The generated electrons flow to the oxic zone through bacterial cables, and the protons flow extracellularly over the pH gradient. The second reaction, in the oxic zone, combines the influx with oxygen to make water. Variant reactions occur. For example, nitrate may be used instead of oxygen as the electron acceptor in the second reaction.

These reactions focus on the flow of electric current. But they leave open three questions. How do cells generate ATP to drive growth? Can cables grow when not connected across habitats? How do single cells grow when alone?

The following discussion fills in background. The background includes significant biochemical detail. The effort to follow that detail is repaid when we arrive at a hypothesis for the cable bacteria life cycle. That life cycle presents a fascinating challenge for understanding microbial life history and the forces that shape metabolic design.

Genomics provides the first piece of the background. Cable bacteria genomes encode diverse metabolic processes. Two studies found genes for ATP-generating pathways on alcohols, hydrogen, and sulfur compounds. The genomes from marine[210] and groundwater[289] habitats differed with regard to the presence of particular genes.

ATP from elemental sulfur and growth of single cells.—Müller et al.[289] developed a hypothesis for how cable bacteria generate ATP from sulfur. Their model arose from culture under different conditions, in which they grew cable bacteria obtained from groundwater.

In one culture, elemental sulfur was the only electron donor source. That growing culture did not favor significant electron flux over cables, suggesting ATP generation and growth in the absence of electron flux.

The culture contained single cells and cables up to several hundred μm in length. As populations grew, they seemed to contain longer cables and more cables relative to single cells. This culture suggests that single cells may grow into populations with small cables.

Growth on elemental sulfur required an additive that scavenged sulfide. When comparing the classic summary of cable bacteria metabolism and electricity in eqns 17.2, it is not immediately clear why the only sulfur-related requirements for growth are adding elemental sulfur, S^0, and removing sulfide, S^{2-}, which may occur as H_2S or HS^-.

Many bacteria, including species closely related to cable bacteria, use S^0 as their electron-donor source of free energy.[105,289] The cable bacteria genome analyzed in this study contained the necessary genes for that elemental sulfur pathway.[289]

For example, an overall transformation may be[105]

$$4\,S^0 + 4\,H_2O \longrightarrow SO_4^{2-} + 3\,HS^- + 5\,H^+. \tag{17.3}$$

Dropping the protons simplifies the reaction to

$$4\,S^0 + 4\,O^{2-} \longrightarrow SO_4^{2-} + 3\,S^{2-}. \tag{17.4}$$

The unusual biochemical aspect arises because the transformation happens by a combination of two coupled redox reactions,

$$S^0 + 4\,O^{2-} \longrightarrow SO_4^{2-} + 6\,e^- \tag{17.5a}$$

$$3\,S^0 + 6\,e^- \longrightarrow 3\,S^{2-}, \tag{17.5b}$$

in which S^0 is the electron donor in the first oxidation reaction and the electron acceptor in the second reduction reaction. This disproportionation of sulfur[105] may be summarized as

$$SO_4^{2-} \longleftarrow 4\,S^0 \longrightarrow 3\,S^{2-}. \tag{17.6}$$

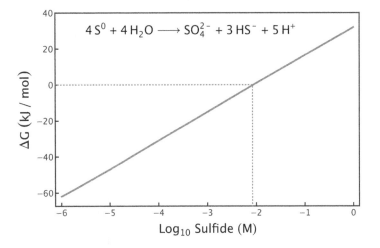

$$4\,S^0 + 4\,H_2O \longrightarrow SO_4^{2-} + 3\,HS^- + 5\,H^+$$

Figure 17.3   Increasing sulfide (HS⁻) concentration stops metabolism of S^0 by product inhibition, raising the free energy change above zero. Dotted line shows the threshold. The reaction summarizes the sulfur-disproportionation pathway in eqns 17.3–17.5. Theoretical calculation at 4 °C and 28 mM SO_4^{2-}. Higher temperature (25 °C) or less sulfate (3 mM) roughly doubles the sulfide threshold for the reaction to proceed. Redrawn from Fig. 1 of Finster.[105]

The left-side oxidation of sulfur to sulfate can produce ATP, providing the free energy disequilibrium to drive growth.

A buildup of sulfide inhibits this metabolic pathway. Figure 17.3 shows that increasing sulfide concentration blocks the reaction.

The sulfur cycle.—One study[289] showed that cable bacteria can grow on S^0. How does that relate to the commonly cited pathway for generating electron flux in eqn 17.2a, which begins by uptake of sulfide, S^{2-}?

Figure 17.4 provides a model for local sulfur cycling, ATP generation, and electron flux over cables.

In that model, all suboxic cells can oxidize sulfide to sulfur, $S^{2-} \longrightarrow S^0 + 2\,e^-$, which does not generate ATP. The electrons flow up the cable to the oxic zone to reduce oxygen to water.

When the sulfide concentration is sufficiently low, suboxic cells disproportionate sulfur into sulfate and sulfide, as in eqn 17.6. The $S^0 \longrightarrow SO_4^{2-}$ pathway generates ATP.

Cells excrete SO_4^{2-}. Sulfate-reducing bacteria are common.[197] They use essentially the same reactions as disproportionation but, compared with

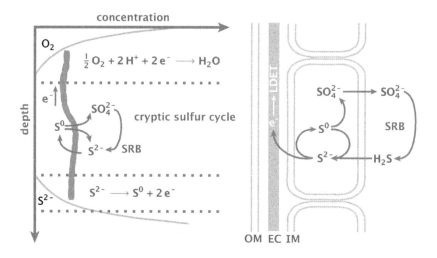

Figure 17.4   Model for sulfur cycling by cable bacteria and sulfate-reducing bacteria (SRB). See text for the steps in various pathways. The cells on the right show the inner membrane (IM), outer membrane (OM), and electron cable (EC) over which long distance electron transport (LDET) runs. Redrawn from Fig. 4 of Müller et al.[289]

eqn 17.6, they run the sulfate-sulfur reaction in reverse,

$$SO_4^{2-} \longrightarrow S^0 \longrightarrow S^{2-}, \tag{17.7}$$

which regenerates sulfide and completes the local cycle.

The overall model in Fig. 17.4 distinguishes three zones. First, the upper zone uses the incoming electrons to reduce oxygen or nitrate.

Second, the middle zone runs a complete sulfur cycle. Cable bacteria take up sulfide and produce sulfate. The sulfate-reducing bacteria transform sulfate into sulfide. Little net change may be observed. That apparent stasis misleadingly suggests limited metabolic activity.

Third, near the anoxic cable ends, sulfide concentration may rise because the cable bacteria do not take it up sufficiently quickly. High sulfide concentration blocks the ATP-generating sulfur disproportionation pathway. Thus, the anoxic ends may act as nongrowing sulfide oxidation uptake terminals, passing their electrons up the cable.

This model completes the summary of metabolism and the life cycle. With that background, I turn to the forces that shape life history.

Habitat niche construction.—In the sulfur-cycle model, high sulfide concentration inhibits the sulfur disproportionation pathway (eqn 17.6) that provides ATP for growth. A nongrowing cable terminal, down in a sulfide-rich zone, could lower the sulfide concentration by oxidation,

$$S^{2-} \longrightarrow S^0 + 2\,e^-,$$

sending the electrons up the cable and raising the nearby S^0 concentration (Fig. 17.4). Once the sulfide concentration drops sufficiently, the pool of S^0 provides food for growth. S^0 may also diffuse upward into the active growth zone, providing additional food to enhance reproduction.

This niche construction[302] process makes the habitat better for growth. A cable that improves the habitat for itself also improves conditions for neighboring cables.

Put another way, lowering the sulfide concentration by electron conduction over the cable provides a shareable public good for neighbors. The neighbors include connected cells attached to the same cable and the many separate cables that likely reside nearby.

The extensive theory for public goods evolution applies. That theory leads to a wide variety of comparative hypotheses in relation to demography and the genetic structure of populations (Chapter 5).

Most simply, greater genetic relatedness typically favors more investment in public goods. In this case, separate cables with more closely related cells are more likely to improve high sulfide environments in a way that benefits all neighbors.

Demographically, long-lived patches typically gain more from niche modifications than short-lived patches.

Tuning of rate versus yield.—Initial studies tend to focus on habitats with high bacterial density. It is easier to find and observe a lot of bacteria than it is to study sparse populations.

High density often associates with genetic mixing. More mixing favors faster growth rate at the expense of reduced yield (Section 4.1).

Isolated resource patches inevitably come and go. Such patches often contain more highly related populations, which tend to favor greater yield and reduced growth rate. Patch lifespans and rates of resource influx also vary, altering the forces that favor rate versus yield.

Demographically, short-lived patches typically favor faster growth. Long-lived patches with limited resources typically favor higher yield.

Those abstract forces apply widely across microbes. It would be particularly interesting to study the mechanisms by which those forces tune the unusual metabolism of cable bacteria.

The scale of competition and dispersal.—Public goods and high yield provide cooperative benefits to neighbors, enhancing group success. High relatedness typically favors those cooperative traits. However, if neighbors also compete for resources, that competition can offset the benefits and disfavor the cooperative traits (p. 72).

Demography plays a role. For example, public goods, such as beneficial niche construction, may affect close neighbors. By contrast, competitive interactions, such as dispersal and colonization of new patches, may happen between distant competitors.

Comparatively, the more intensely cooperation happens locally between close relatives and the more intensely distant competition happens globally against unrelated individuals, the more strongly natural selection will favor cooperative traits.[122]

Spatial scaling of cooperation and competition depends on movement. Cables actively move, as summarized above. Movement may primarily serve to maintain electron flux between anoxic and oxic zones.

Additional movement in excess of what is needed to maintain electron flux may occur in order to increase dispersal distance, reducing competition between close relatives. Comparatively, more local competition between relatives favors greater excess dispersal.[111,169]

Budding and breakpoints in the formation of new cables.—As cables grow, they must eventually break into smaller pieces. Those buds are the cables' progeny.[150]

A long cable could produce many single cells and short cables. Or it could break near the middle, making fewer large buds. Demographically, the number of offspring trades off against the size per offspring.[387]

Changing conditions alter the relative reproductive value of small and large offspring (Section 5.6). For example, short-lived resource patches favor dispersal, which likely values many small progeny more highly than a few large progeny.

By contrast, a strong growth premium for cables that connect anoxic to oxic zones likely values a few long progeny over many short ones.

Mechanistically, the breakpoint tendency between various cells may evolve in response to changing design forces on progeny size.

Many aspects of resource distribution, demography, and genetic structure influence the design forces that favor different-sized offspring. I leave the development of comparative predictions as an open challenge.

Hookups between cables.—Can a suboxic cable dump excess electrons by physically connecting to a cable with an oxic terminal? There are a few clues but no direct evidence.

Reimers et al.[338] placed a carbon electron-accepting anode into river estuary sediment. After 412 days, they found cable bacteria filaments of more than 40 cells firmly attached to the electrode. Presumably, the short filaments in the anoxic habitat dumped their excess electrons through their physical connection to the electrode.

Competitive habitats may favor hookups between cables. Linking cables increases the rate at which short filaments can span zones to transmit electrons. Interspecies links may bring synergistic metabolic pathways into proximity, allowing joint exploitation of new resources.[252]

Extracellular Electron Shuttles

Other microbes transmit electrons extracellularly. This subsection focuses on shuttle molecules that pick up electrons at the cell surface, move by diffusion, and dump the electrons on a distant recipient.

Those freely diffusing shuttle molecules can potentially be used by any nearby cells. Some cells may not make any shuttles, instead using the shuttles made by other cells.

The public goods problem arises when a cell produces a shareable resource that diffuses away and can be used by any neighbor (Chapter 5). The producer pays the cost. All neighbors can benefit.

The overall costs and benefits of diffusible shuttles depend on the molecular mechanisms, the similarity or relatedness between neighbors, and demography.

Thermodynamic background.—Catabolic cascades move electrons from donor molecules to recipient molecules. Donor molecules provide food. Recipient molecules provide an electron sink.

Electrons come from food, pass through the catabolic cascade, and must be dumped after use. The thermodynamic driving force depends on electron flux through the full path. Cells fuel life by capturing free energy from that electron flux.

For microbes that can access and use oxygen, finding an electron sink provides little challenge. For microbes in anaerobic conditions, accessing a good electron sink may be just as challenging as obtaining food.

In some cases, the internal catabolic cascade provides the final electron sink. For example, the molecules that accept electrons to make acetate may be the final electron sink for anaerobic fermentation.

Often a cell must dump the fermentation product to prevent product inhibition. In essence, the overflowing cells dump the recipient electrons to keep the sink open for further electron flow.

Extracellular electron sinks.—Cable bacteria in anaerobic habitats reach an oxygen electron sink by aggregating to make a wired link. Electrons flow over the cellular wires to the distant oxygen source.

Microbes use a variety of extracellular mechanisms to dump electrons to various sinks[217,252,376] (Fig. 14.1). Some species make extracellular electric pili. Those pili pass electrons directly to inorganic electron sinks or to cells of other species.

Electric pili have been studied mostly in *Geobacter*. Conducting pili have been observed in several other genera. Homologous genes for electric pili are widely distributed throughout Archaea and Bacteria.[253]

Other mechanisms of direct electron transfer by physical contact have been documented in many species.[252,458] For example, an electron shuttle sequence moves electrons between the inner membrane and the thick outer cell wall of gram-positive *Thermincola*. Surface cytochromes on the cell wall can transfer electrons extracellularly.

Diffusible extracellular shuttles.—Cells could potentially dump electrons to various sinks. However, many potential electron sinks are insoluble and occur at a distance from cells.

For example, insoluble forms of ferric iron, Fe^{3+}, occur widely.[203] Under many conditions, ferric iron takes up electrons to make the reduced ferrous form, Fe^{2+}.

Diffusible extracellular shuttles can transport electrons from cell surfaces to distant Fe^{3+} or other electron sinks. Most studies use human-placed electrodes as electron sinks.[217,252,376,458]

Experimentally manipulating diffusible electron transport alters catabolism, survival, or growth. Manipulations include introducing the electrode, adding potential shuttles to the medium, or replacing the medium with fewer potential shuttles.

Genes associated with extracellular electron transport occur widely among microbes, suggesting broad functional significance.[242,276] Genetic manipulations and knockouts of putative shuttle-pathway components provide direct evidence for the function of particular molecules.[155,213,242,243]

Extracellular shuttles have multiple functions.[195,203] In addition to dumping electrons to enhance catabolism, extracellular electrons can also enhance cellular iron scavenging by reducing insoluble ferric iron to soluble ferrous iron, $Fe^{3+} \longrightarrow Fe^{2+}$.

Siderophore analogy.—Microbes often produce siderophores, an alternative type of extracellular shuttle to scavenge iron.[216,235]

We know much more about siderophores than about electron shuttles. Analogies with siderophores provide the best way to develop comparative predictions for electron shuttles.

I outline the forces that act on siderophore design. I then relate the siderophore analogies to prospects for studying electron shuttles.

Siderophore biology.—Iron often limits microbial growth.[15] The metal mostly occurs in the insoluble ferric form. Cells compete for soluble ferrous iron, which often remains too rare to satisfy demand.

To obtain the additional required iron, microbes secrete siderophores. Those extracellular iron shuttles can bind insoluble Fe^{3+}. Cells take up the diffusible iron complexes.

Individual cells produce siderophores. Free siderophores can be used by any neighboring cell. A user requires the matching receptor to bind free siderophores and the associated pathways to take up the iron.

Nonproducing cells often express receptors for several different siderophore specificities, including those made by other species. Producing cells may also take up variant siderophores made by others.

Individual production and shared use define public goods. Making a public good is often described as a cooperative trait because it enhances neighbors' success.

Nonproducers may loosely be described as competitive cheaters because they use their neighbors' products without themselves contributing to group success.

Several classic forces of design shape siderophore traits. The following summarizes a few of those forces. Genetic structure and the diversifying forces of competition play particularly important roles.

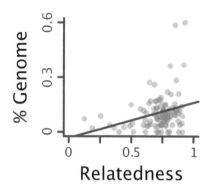

Figure 17.5 Percentage of genome coding for siderophore-related genes. Circles represent 101 bacterial species from 239 human stool samples. Relatedness measures genetic similarity within human hosts relative to the population of hosts. Redrawn from Fig. 3 of Simonet & McNally.[384]

Genetic structure.—In terms of genetic transmission to the future, a public good has the same beneficial effect whether it returns to the original producer or to a genetically similar neighbor. Thus, greater relatedness typically favors more production of public goods.

Simonet & McNally[384] rephrased this prediction by using genomic coding for siderophores as a proxy for public goods production. Species with a greater tendency to live near genetic neighbors devote more of their genome to siderophores. Figure 17.5 supports that prediction.

Comparison: matching the change in force to the change in traits.— Classic comparative studies contrast traits between species or higher-level taxa.[173] Ideally, one compares each past environmental change with the associated divergence in species' characters.

I have used the same logic throughout this book. In particular,

$$\text{parameter} \rightarrow \text{force} \rightarrow \text{trait},$$

a change in an environmental parameter changes a fundamental force, which changes a trait. Ideally, one matches the timescales over which environments and traits change.

Relatedness measures the genetic structure of a population. I have often mentioned varying relatedness as a consequence of changing environmental parameters. A shift in relatedness alters the fundamental force of kin selection (Section 5.2).

Relatedness likely changes over short temporal and spatial scales. By contrast, genomes likely change more slowly than relatedness. Thus, comparing short-term estimates of species-level genetic relatedness to long-term changes in genomic attributes provides an interesting but rather weak analysis (Fig. 17.5).

In other cases, environmental change may happen slowly on the same timescale as species differences. For example, broad changes in environmental habitats and the evolved capacity to live in those different habitats may change relatively slowly.

For many microbial traits, the evolutionary tuning of expression may happen quickly in response to environmental changes. For example, expressing more or less of a public good may evolve rapidly.

Microbes provide unique opportunities to study short timescales for both environmental and evolutionary change. Matched timescales best reveal how fundamental evolutionary forces shape design.

Alternative timescales expose different aspects of design. However, the longer the timescale, the more difficult it becomes to trace pathways of partial causation from changed environments to changed traits.

Nonproducers arising from cellular heterogeneity in vigor.—In theory, vigorous cells pay a smaller marginal cost for secretion than weak cells because an incremental cost forms a smaller fraction of a large resource pool than a small resource pool (Section 5.5).

Comparatively, increasing vigor reduces marginal costs. Lower marginal costs raise secretion. Weak cellular vigor may associate with high relative costs and nonproduction.

Greater heterogeneity in vigor between cells increases the heterogeneity in marginal costs of production, which predicts greater heterogeneity in cellular secretion. Higher cellular heterogeneity in vigor may associate with more nonproducing cells.

Diversifying forces of competition.—Cells often gain by using the siderophores produced by others.[216] Taking up foreign siderophores requires the matching receptor.[179,334]

A producer can limit foreign usage by making a private receptor-binding motif. Privacy favors usurpers to evolve new matching receptors. Continual conflict between private usage and usurpation diversifies siderophore specificities.[55,137,295]

Several species express diverse siderophores and matching receptors. Some species express diverse receptors to take up siderophores made by other species.[19,60,68,69,178]

Siderophore receptors provide a site of attack by bacteriophage and toxic bacteriocins.[179] Attack imposes an alternative force favoring receptor diversity.

Comparatively, greater mixing of genotypes or species favors more cross-type uptake and enhanced diversity. The habitat's biophysics affects the mixing of types and the pressure to diversify.

Similarly, greater intensity of attack by bacteriophage and bacteriocins favors greater receptor diversity. The spatial scale of movement influences the intensity and diversity of attack.

Biophysics may dominate.—Diffusion sets spatial scale, influencing many aspects of siderophore biology.[149,218,235,275]

Secreted siderophores benefit a local group only when the siderophores do not diffuse away too rapidly. Nonproducers gain by taking up others' siderophores only when the siderophores diffuse sufficiently rapidly to reach the nonproducers.

Cells may mix on a different spatial scale from the siderophores. For example, cells may attach to surfaces but interface with a high-diffusion medium. Or a viscous medium may cause different movement patterns by the differently sized siderophores and cells.

The challenge of partial causation.—How does increased diffusion rate alter siderophore traits? The overall effect depends on the interaction between several pathways of partial causation.

Greater diffusion may mix cells, shrinking the spatial scale of genetically related neighbors. Changes in relatedness can strongly influence the benefits of secreting diffusible siderophores. Diffusion can also alter the spatial scale of competition for resources.

In addition, greater diffusion may increase the opportunity to use siderophores produced by others. Nonproducers may increase.

Media that reduce diffusion favor biophysical modifications of siderophores to increase their diffusion rate. Diffusive media favor siderophores that remain attached to the cell surface. Observations support those associations between biophysics and siderophore traits.[218]

When analyzing siderophores, several studies focus solely on genetic relatedness, competition, and spatial scaling. Those generic fundamental forces broadly influence traits in many different circumstances.

Other studies focus on biophysical processes, such as diffusion. Altered diffusion changes genetic relatedness, which shapes siderophore expression level. In a different partial causal path, varying diffusivity of media alters biophysical properties of siderophores, which may in turn modify spatial scale, relatedness, and demography.

Useful explanations arise by parsing the individual partial pathways and then analyzing how those pathways shift in dominance and interact.

Design forces for catabolism via diffusible shuttles.—We know less about electron shuttles than about siderophores. Preliminary data suggest similarities and the potential to develop analogous comparative predictions.

For both shuttles and siderophores, some cells release extracellular vehicles. Neighboring cells can reuse those vehicles. Both types of vehicles target ferric iron, although shuttles also target other sources.

Shuttles may enhance publicly usable iron when dumping an electron to reduce insoluble ferric iron to soluble ferrous iron, $Fe^{3+} \longrightarrow Fe^{2+}$.

Differences occur. Habitat structure and target distribution affect electron sinks differently from iron acquisition. Electrochemical demands differ, likely altering how biophysical properties influence costs, diffusibility, local density, and reuse.

We lack the background information needed to develop compelling comparative predictions for electron shuttles. I briefly summarize some additional facts from preliminary studies.

Public use, private use, and diversity of shuttles.—The foodborne pathogen *Listeria monocytogenes* can use flavin shuttles to carry electrons to extracellular ferric iron sinks. Under some laboratory conditions, growth requires extracellular electron shuttling.[242,243,353]

The primary natural function of extracellular iron reduction remains unclear. It may function in catabolism, iron scavenging, or both.[195] Genomes of 31,910 prokaryotic genomes show widespread distribution of genes involved in flavin-mediated extracellular electron functions.[276]

Listeria monocytogenes cannot produce flavin shuttles for extracellular electron transport.[242] In its natural environment, sufficient extracellular flavins commonly occur.[187,328] Public use of flavins may be common.

Uptake by nonproducers may favor private shuttle-receptor pairs. I did not find any relevant studies.

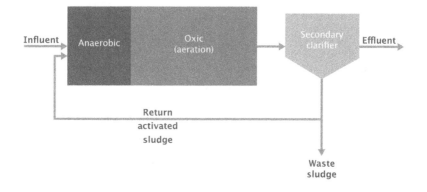

Figure 17.6 Biological phosphorus removal process. Alternating anaerobic and aerobic periods influence the biochemical transformations by bacterial species. Shown previously as Fig. 15.1, where the text provided details. Redrawn from Fig. 17 of Curtin et al.[72]

17.3   Storage When Resources Fluctuate

Tradeoffs arise between the current use of free energy and the storage of free energy for later use. Microbial wastewater treatment illustrates those tradeoffs.

Wastewater Treatment: Demography and Storage

An earlier section introduced a particular microbial wastewater treatment cycle (p. 231). Here, I briefly review the cycle. I then use the demography as a model challenge in the study of design (Section 5.6).

Figure 17.6 summarizes the cycle. The influent contains phosphorus and organic carbon waste. Alternating anaerobic and aerobic bacterial processing cleans the influent.

The engineering challenge seeks the optimal sequence of environments. However, once the treatment begins, subsequent uncontrolled bacterial change may limit success.

In the treatment cycle, ecological processes favor colonization and extinction by different bacteria. Evolutionary processes alter the metabolic flux of particular species.

The treatment parameters induce demographic forces, which favor certain microbial traits over others. The ideal treatment cycle optimizes

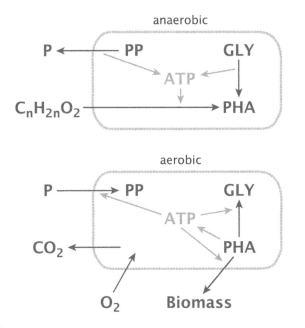

Figure 17.7 Bacterial transformations in wastewater treatment. The transformations summarize the composite changes by various bacterial strains and species. Abbreviations: short-chain fatty acids ($C_nH_{2n}O_2$), phosphate (P), polyphosphate storage (PP), glycogen storage (GLY), and polyhydroxyalkanoate storage (PHA). These drawings simplify the more complete descriptions in Figs. 15.2 and 15.3.

the initial bacterial composition and the subsequent ecological and evolutionary changes.

Figure 17.7 shows the key biochemical transformations and bacterial traits. With regard to engineering goals, the waste influent starts with phosphate (P) and organic carbon in various short-chain fatty acid fermentation products ($C_nH_{2n}O_2$), such as formate, acetate, or butyrate.

In the anaerobic phase, bacteria lack an electron acceptor to catabolize the fatty acids. Instead, they use internal polyphosphate (PP) or glycogen (GLY) to generate ATP. That free energy drives the transformation of fatty acids into polyhydroxyalkanoate (PHA) storage.

The aerobic phase uses the PHA store to drive biomass synthesis and to rebuild the PP or GLY stores.

From an engineering perspective, the net transformations oxidize organic carbon to CO_2 and bind external P into cellular PP stores. Discarding the residual bacteria as waste sludge cleans the water.

One problem concerns GLY versus PP storage. In the aerobic phase of Fig. 17.7, limited resources impose a tradeoff between making GLY or PP. If dominant strains tend to build more GLY, then the system does not clear P contaminants.

In practice, GLY-favoring strains sometimes do increase.[351,383] Apparently, those GLY-favoring strains gain survival or growth advantages over the life cycle.

What sort of environmental changes would enhance the relative fitness of the beneficial PP-favoring strains? That question demands more detailed information than we have at present.

To set a foundation for future work, I briefly consider how fitness arises from the interaction between demographic and physiological tradeoffs. I do so in an abstract way, relating demography to general forces of design.

DEMOGRAPHY AND THE FORCES OF DESIGN

The life cycle has two generic aspects. First, cells pass through alternative habitats. Second, internal stores drive growth in one habitat and drive other fitness components in the alternative habitat.

I illustrate how to analyze simple forces of design in a complex life cycle. I do not try to match the particular biochemical dynamics shown in Fig. 17.7. Instead, I consider abstractly how internal cellular storage may influence growth and survival in two different habitats.

After illustrating the analysis of generic forces, I discuss how these methods may be applied to complex life cycles. Such life cycles must be common in nature but remain difficult to study.

Industrial microbiology provides many examples of complex life cycles. Often, the uncontrolled evolutionary response of microbes sets a primary challenge for successful application. Such systems are good models for studying the forces of design.

I use the word *evolutionary* in a broad way. In nonrecombining microbes, every distinct genotype competes as a variant. Some variants may be mutant strains of a species. Other variants may be distinct species. The system evolves by the changing abundances of the variants.

Growth, storage, and survival in two habitats.—Suppose microbes encounter two different environmental conditions. In the first habitat, H_1,

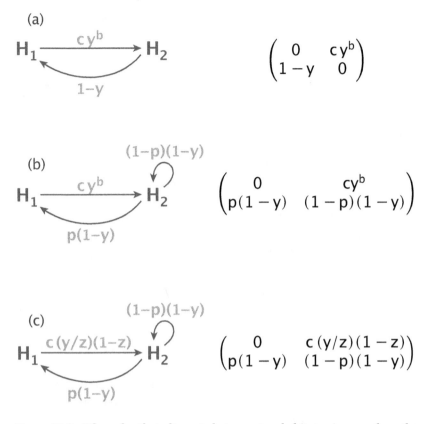

Figure 17.8 Life cycles that alternate between two habitats. Arrows show the fitness contribution of each habitat to subsequent habitats. The **A** matrices on the right summarize the fitness contributions, in which each entry, w_{ij} represents the contribution of habitat j to habitat i.

they use a fraction y of their resources for growth and a fraction $1 - y$ to build intracellular storage. In the second habitat, H_2, survival depends on the intracellular store.

Figure 17.8a shows a life cycle that alternates between the two habitats. In habitat H_1, allocating y resources to growth increases the cellular population by cy^b, in which c and b are parameters.

Because the life cycle alternates between habitats, the full cycle fitness is the product of growth in H_1 and survival in H_2, thus $w = cy^b(1 - y)$. The optimal allocation is $z^* = b/(b + 1)$, which is obtained from the standard maximization procedure of evaluating $dw/dy = 0$ at $y = z^*$.

The simple life cycle in Fig. 17.8a allows us to write the fitness expression directly. For the more complex life cycles in Fig. 17.8, it helps to evaluate components of fitness in relation to demography.

Demography and reproductive value.—We can use life cycles in Fig. 17.8 to illustrate demographic analysis (Section 5.6). Those methods highlight general forces of design that apply to complex life cycles.

Equation 5.10, which expresses fitness within its full demographic context, is repeated here,

$$W = \sum_{ij} v_i w_{ij} u_j = \mathbf{v}\mathbf{A}\mathbf{u}. \tag{17.8}$$

In the matrix **A**, each element w_{ij} defines the fitness contribution of habitat j to habitat i. The matrices in Fig. 17.8 show **A** for each scenario.

The column vector **u** gives the fraction of the total population in each habitat. The greater the population size in a habitat, the more that habitat contributes to the future of the overall population.

The values in **u** influence the overall design force that shapes traits. That overall force combines the force in each habitat weighted by the population size in that habitat. A force that acts locally on relatively few individuals contributes little to the overall force.

The row vector **v** gives the reproductive value of each individual in each habitat. Reproductive value describes the expected contribution of an individual to the future of the population.

An individual in a poor habitat typically contributes relatively little, whereas an individual in a good habitat typically contributes relatively more. Forces acting on individuals with low reproductive value matter less than forces acting on individuals with high reproductive value.

We can analyze the various scenarios in Fig. 17.8 by noting that all share the same form of the fitness matrix as

$$\mathbf{A} = \begin{pmatrix} 0 & n \\ s & t \end{pmatrix}, \tag{17.9}$$

in which each entry is a function of individual trait value, y and, in some cases, also the average trait value, z, of a local group. Optimal trait values at equilibrium occur at $y = z = z^*$.

The population grows at rate λ, the dominant eigenvalue of **A**, obtained by solving

$$\lambda^2 - t\lambda - sn = 0,$$

in which we evaluate the terms at their equilibrium trait values, z^*.

From Frank [122, p. 147], the equilibrium habitat population sizes and individual reproductive value weightings are proportional to

$$\mathbf{u} \propto \begin{pmatrix} n^* & \lambda \end{pmatrix}$$
$$\mathbf{v} \propto \begin{pmatrix} s^* & \lambda \end{pmatrix},$$

showing the column vector $\mathbf{u}$ as a row. I use the '*' superscripts to emphasize that this analysis measures the demographic context at equilibrium, $y = z = z^*$. I comment below on the equilibrium assumption.

The overall fitness expression in eqn 17.8 divided by λ^2 is

$$W = (sn^* + s^*n)/\lambda + t. \tag{17.10}$$

The unstarred functions, s, n, and t, depend on variable trait values, y and z.

This fitness expression accounts for the way in which genes pass through various habitats over the life cycle. For example, the lower diagonal element of $\mathbf{A}$, which is t, contributes directly to the same habitat in the next time step.

By contrast, the off-diagonal elements, s and n, send genes through the alternative habitat. For example, s accounts for the fitness effect of variable traits in the first habitat with respect to transmission to the second habitat.

The second habitat then multiplies by n^* the initial transmission from the first habitat, in which we take the future multiplications at the equilibrium value. Thus, we can think of trait variation as causing an instantaneous perturbation of force. We then follow that perturbing force through the life cycle in its equilibrium context.

To transit through the cycle, the two-step paths for the off-diagonal elements require one more step than the on-diagonal elements. Thus, with an extra time step, the two-step contributions happen as the population has grown by an additional amount, λ. Those two-step contributions must therefore be devalued by the population expansion factor.

The two off-diagonal entries experience the same devaluation relative to the on-diagonal element. Thus, the term $(sn^* + s^*n)/\lambda$ describes the valuation of the two-step paths relative to the one-step path, t. In more complex life cycles, the method tracks the pathways through various environments, weighting each path by its relative contribution to the future population.

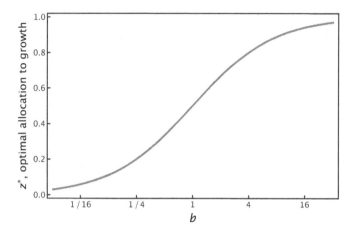

Figure 17.9   Fraction of resources allocated to growth, z^*, with the remaining fraction $1 - z^*$ allocated to storage and survival. Based on the life cycle in Fig. 17.8a, in which individuals pass alternately through growth-favoring and survival-demanding habitats. The optimal allocation depends on b, which determines the marginal gains for growth relative to survival, leading to the solution in eqn 17.13.

Forces acting on traits.—We evaluate forces by studying how changes in traits alter fitness. If we take the demographic context at equilibrium, then we have the fitness expression, W, in eqn 17.10.

As an individual makes small changes in its trait, y, fitness changes by dW/dy evaluated at demographic equilibrium and at trait value equilibrium, $y = z = z^*$, yielding

$$\frac{dW}{dy} = (s'n^* + s^*n')/\lambda + t',$$

in which primes denote differentiation with respect to y evaluated at trait value equilibrium. We can find a candidate optimum by setting this expression to zero. Multiplying by λ/s^*n^* yields

$$\frac{s'}{s^*} + \frac{n'}{n^*} + \frac{t'}{s^*n^*/\lambda} = 0. \tag{17.11}$$

At the optimum, for each fitness component, any normalized marginal gain with respect to changing trait value is balanced by an equal marginal loss in the other fitness components. These marginal changes follow from the force perturbations caused by trait variations near equilibrium.

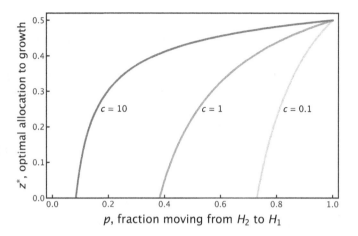

Figure 17.10 Optimal growth allocation in relation to p, which determines the fraction of the life cycle spent in survival-demanding (H_2) versus growth-favoring (H_1) conditions. A rise in c increases the relative efficiency of converting resources into growth versus storage. Based on the life cycle in Fig. 17.8b and the solution in eqn 17.14.

Marginal growth benefit.—For the case in Fig. 17.8a, the marginal valuations at the optimum are

$$\frac{n'}{n*} = -\frac{s'}{s*},$$

which yields

$$\frac{b}{z*} = \frac{1}{1 - z*}, \tag{17.12}$$

leading to

$$z* = b/(b + 1) \tag{17.13}$$

for the optimal fraction of resources allocated to growth in H_1 instead of survival in H_2. The marginal benefit for growth rises with b, favoring more investment in growth, $z*$, and less in survival, $1 - z*$ (Fig. 17.9).

Repeated survival challenge and reproductive value.—The life cycle in Fig. 17.8b splits survivors from the nonreproductive habitat, H_2, into a fraction p that return to the reproductive habitat and a fraction $1 - p$ that remain in H_2. When remaining in H_2, survival depends on continued use of stored resources.

The extended periods in H_2 raise the valuation of survival relative to growth. Demographic analysis provides a way to study the relative values of alternative fitness components.

Using the general solution in eqn 17.11 and assuming $b = 1$ to highlight the demographic factors in the life cycle, we obtain the optimum fraction of resources allocated to growth instead of storage and survival as

$$z^* = \frac{1 - p - \sqrt{cp}}{1 - p - 2\sqrt{cp}}. \tag{17.14}$$

Figure 17.10 illustrates the solution. As p declines, individuals increasingly remain in H_2, where they must survive on their stored resources. That greater valuation weighting for survival decreases the optimal fraction of resources allocated to growth.

By contrast, an increase in c enhances the relative efficiency of converting resources to growth versus storage, favoring greater allocation to growth.

Rate versus yield and the genetic structure of populations.—For the life cycle in Fig. 17.8c, growth in habitat H_1 includes competition between different genotypes. Fitness in that habitat follows the basic tragedy of the commons scenario from eqn 5.1. When $c = 1$ we have

$$w = \frac{y}{z}(1 - z),$$

in which y is a random focal individual's fractional allocation to growth, and z is the local group of competitors' average allocation to growth. Using the methods in Section 5.2, the optimal allocation to growth when applied to this fitness component alone would be

$$z^* = 1 - r,$$

in which r is a generalized notion of a kin selection coefficient as $r = dz/dy$, the slope group phenotype on the focal individual's phenotypic value that transmits to future generations. At the optimum, with $y = z = z^*$, fitness is $w = r$.

The optimum expresses a simple rate versus yield tradeoff. As relatedness, r, increases, the allocation to growth, z^*, declines. With less allocation to growth, the competition between different types for local resources also declines.

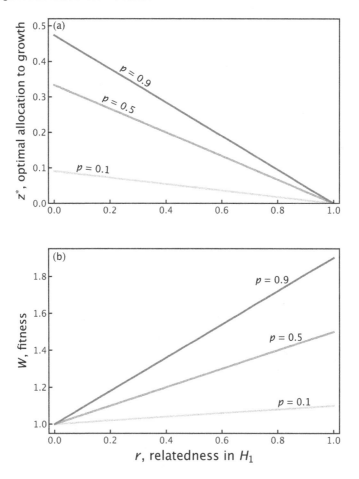

Figure 17.11   Mixed-type competition in the growth habitat, H_1, induces a growth rate versus biomass yield tradeoff that alters the growth versus storage allocation. As the parameter r rises, less mixing occurs, favoring greater overall yield efficiency and less allocation to growth. Based on the life cycle in Fig. 17.8c and the solution in eqn 17.15.

Lower short-term growth rate and resource competition increase fitness, effectively raising the yield per unit of available resource.

Alternatively, lower relatedness causes more competition between types, faster short-term growth, and lower efficiency and long-term yield.

In this life cycle, the tragedy of the commons in H_1 embeds within a broader demographic context. For the full demography in Fig. 17.8c, and with $c = 1$ for simplicity, the optimum is

$$z^* = \left(\frac{p}{1 + p}\right)(1 - r),   \tag{17.15}$$

with fitness at the optimum of $W = 1 + pr$, illustrated in Fig. 17.11.

Demographically, lower p values increase the time that individuals spend in the survival-demanding habitat, H_2, favoring greater allocation to storage over growth.

APPLICATIONS TO INDUSTRIAL MICROBIOLOGY

In the wastewater treatment cycle of Fig. 17.6, controlling microbial competition and evolution sets a primary challenge in application. Many applied microbiology problems face similar challenges of managing microbial evolution.

How can we connect the simple abstract models to those real problems and their complex life cycles? That question defines a significant and challenging unsolved research goal.

One often gains the most insight by analyzing the same problem with multiple analytical and conceptual approaches. If attacking this challenge myself, I would consider initially a classical differential equation analysis and an agent-based computer simulation, allowing all potentially interesting assumptions about the biology.

I would then try to parse the complex dynamics into understandable forces. How do those forces dominate as causes of variation in outcome? Can I formulate partial causes that can be expressed as testable comparative hypotheses? Almost always, progress in understanding will depend on simple, clear comparative hypotheses.

It may be possible to improve control by analyzing dynamical models of how populations change. However, such control by dynamical analysis alone is often achieved without a clear understanding of the underlying forces that drive the change. Without that understanding, solutions tend to be fragile. By contrast, understanding the forces allows us to predict how an altered force changes motion.

Industrial microbiology provides great opportunities for the study of microbial design. Progress in understanding the forces of design will improve efficiency in application.

17.4 Challenges in the Study of Design

The simple models we have discussed have strengths and weaknesses.

On the strong side, the models highlight the three fundamental forces of design.[122] Marginal values set the tradeoffs between alternative allocations for limited resources. Reproductive values account for how demography weights fitness components, such as growth, survival, and success in different habitats. Generalized kin and similarity selection measure how interactions between different types alter the transmission of traits through time.

These fundamental forces recur in essentially every natural situation. Changes in the environment often influence design by altering these fundamental forces.

For example, biophysical properties such as diffusion of resources or viscosity-limited cellular movement may have several consequences. They can alter the marginal gains obtained in return for resources allocated to different traits, modify the demography of the population, and change the amount of mixing between genetic variants.

On the weak side, these simple models face various difficulties. Optimal traits in equilibrium rarely exist in nature. In any particular situation, many unmodeled environmental parameters alter trait values and population dynamics. The list goes on.

Why have I emphasized simple models, given their certain failure to explain exactly what we see in any application?

Because this book is about isolating component forces and partial causes. Studying how natural processes design organisms is profoundly difficult. We have no chance unless we can break down the complexity into smaller pieces that we can potentially study and understand.

Simple models, with a focus on fundamental forces, provide one tool to study design. Other analyses, such as standard models of differential equations and dynamics, provide different tools.

When building something complex, we require multiple tools. To say that simple models are misleading because they are not sufficient by themselves is like saying that a screwdriver will disrupt making a house because you cannot build a house with just a screwdriver.

In the simple models, I emphasized component forces. Those component forces isolate partial causes. We understand design by building up understanding from partial causes.

To make progress on real systems, we often need a combination of tools. Standard dynamical models and agent-based computer simulations allow description of detailed natural history and biophysics, with a connection to measured parameters. One gets a sense of how things change. The moving body tells the story of evolutionary design.

The simple models analyze the fundamental forces that act implicitly in dynamical models. As Lanczos[222] said, to enhance understanding of causes, it often pays "to focus on the forces, not on the moving body." The plot revealed.

Ultimately, the deepest understanding comes from combining the alternative perspectives of motion and force into a unified narrative.

18 Design Revisited

To understand biological design, general principles must meet the facts of nature. Part 1 gave the principles, Part 2 the facts.

Only in Chapters 16 and 17 did I start on joining the two sides. Given the divide that remains between principles and application, why did I write this book? Because building up the two sides highlights the divide. A good failure clears the way forward.

The way forward begins with a method for how to study organismal design. How do we join general principles about the forces of design to the constraining biophysics and facts of nature?

No norms specify the proper method across all biological disciplines. Comparative predictions sometimes occur. Yet, in many fields, that comparative approach applies inconsistently, implicitly, or not at all.

Every comment about design should derive from a comparative hypothesis. As conditions change, we expect traits to change in a predicted way. Inferred causality arises from comparing altered conditions to changed traits, revealing the forces of design.

This book discussed many difficulties with this approach. But there is no better way.

Traditional phylogenetic methods of comparison between species and higher taxa have always had the right idea.[75,173] However, those broad taxonomic scales of observed change often do not match the smaller scales over which conditions change and forces of design act.

Microbes provide the opportunity to match the scales of change to the underlying scales of force. Rapid generations, large populations, and great diversity at all scales allow wide choice in focal points for study.

Comments about design often arise as add-ons to studies done for other reasons. This book emphasized the failure of haphazard inference. Instead, the complexity of biological design demands a disciplined approach matched to the empirical problem.

References

1. Acevedo, M. A. et al. 2019. Virulence-driven trade-offs in disease transmission: A meta-analysis. *Evolution* 73, 636–647.

2. Ackermann, M., Chao, L., Bergstrom, C. T. & Doebeli, M. 2007. On the evolutionary origin of aging. *Aging Cell* 6, 235–244.

3. Ackermann, M., Stearns, S. C. & Jenal, U. 2003. Senescence in a bacterium with asymmetric division. *Science* 300, 1920.

4. Aidelberg, G. et al. 2014. Hierarchy of non-glucose sugars in *Escherichia coli*. *BMC Systems Biology* 8, 133.

5. Alberti, S., Gladfelter, A. & Mittag, T. 2019. Considerations and challenges in studying liquid-liquid phase separation and biomolecular condensates. *Cell* 176, 419–434.

6. Alexander, R. D. 1974. The evolution of social behavior. *Annual Review of Ecology and Systematics* 5, 325–383.

7. Alexander, R. D. 1987. *The Biology of Moral Systems*. New York: Aldine de Gruyter.

8. Aller, R. C. et al. 2019. Worm tubes as conduits for the electrogenic microbial grid in marine sediments. *Science Advances* 5, eaaw3651.

9. Alon, U. 2019. *An Introduction to Systems Biology: Design Principles of Biological Circuits*. 2nd ed. Boca Raton, Florida: CRC Press.

10. Amend, J. P. & Shock, E. L. 2001. Energetics of overall metabolic reactions of thermophilic and hyperthermophilic Archaea and Bacteria. *FEMS Microbiology Reviews* 25, 175–243.

11. Ammar, E. M., Wang, X. & Rao, C. V. 2018. Regulation of metabolism in *Escherichia coli* during growth on mixtures of the non-glucose sugars: arabinose, lactose, and xylose. *Scientific Reports* 8, 1–11.

12. Anand, A. et al. 2019. Adaptive evolution reveals a tradeoff between growth rate and oxidative stress during naphthoquinone-based aerobic respiration. *Proceedings of the National Academy of Sciences* 116, 25287–25292.

13. Anderson, B. D. O. & Moore, J. B. 1989. *Optimal Control: Linear Quadratic Methods*. Englewood Cliffs, New Jersey: Prentice Hall.

14. Anderson, R. M. & May, R. M. 1982. Coevolution of hosts and parasites. *Parasitology* 85, 411–426.

15. Andrews, S. C., Robinson, A. K. & Rodríguez-Quiñones, F. 2003. Bacterial iron homeostasis. *FEMS Microbiology Reviews* 27, 215–237.

16. Andrews, S. S. & Bray, D. 2004. Stochastic simulation of chemical reactions with spatial resolution and single molecule detail. *Physical Biology* 1, 137–151.

17. Åström, K. J. & Murray, R. M. 2008. *Feedback Systems: An Introduction for Scientists and Engineers.* Version v2.11a. Princeton, NJ: Princeton University Press.

18. Atkins, P. W. 1994. *The 2nd Law: Energy, Chaos, and Form.* 2nd ed. New York: W. H. Freeman.

19. Baars, O., Zhang, X., Morel, F. M. M. & Seyedsayamdost, M. R. 2016. The siderophore metabolome of *Azotobacter vinelandii. Applied and Environmental Microbiology* 82, 27–39.

20. Bachmann, H., Molenaar, D., Branco dos Santos, F. & Teusink, B. 2017. Experimental evolution and the adjustment of metabolic strategies in lactic acid bacteria. *FEMS Microbiology Reviews* 41, S201–S219.

21. Bachmann, H. et al. 2013. Availability of public goods shapes the evolution of competing metabolic strategies. *Proceedings of the National Academy of Sciences* 110, 14302–14307.

22. Baker, J. L. et al. 2019. *Klebsiella* and *Providencia* emerge as lone survivors following long-term starvation of oral microbiota. *Proceedings of the National Academy of Sciences* 116, 8499–8504.

23. Baldwin, J. M. 1896. A new factor in evolution. *American Naturalist* 30, 441–451.

24. Balleza, E. et al. 2009. Regulation by transcription factors in bacteria: beyond description. *FEMS Microbiology Reviews* 33, 133–151.

25. Banani, S. F. et al. 2016. Compositional control of phase-separated cellular bodies. *Cell* 166, 651–663.

26. Bar-Even, A., Flamholz, A., Noor, E. & Milo, R. 2012. Rethinking glycolysis: on the biochemical logic of metabolic pathways. *Nature Chemical Biology* 8, 509–517.

27. Basan, M. 2018. Resource allocation and metabolism: the search for governing principles. *Current Opinion in Microbiology* 45, 77-83.

28. Basan, M. et al. 2015. Overflow metabolism in *Escherichia coli* results from efficient proteome allocation. *Nature* 528, 99-104.

29. Basan, M. et al. 2020. A universal trade-off between growth and lag in fluctuating environments. *Nature* 584, 470-474.

30. Beckett, S. 2014. *Nohow On: Company, Ill Seen Ill Said, and Worstward Ho.* New York: Grove/Atlantic.

31. Begg, K. J. & Donachie, W. D. 1985. Cell shape and division in *Escherichia coli:* experiments with shape and division mutants. *Journal of Bacteriology* 163, 615-622.

32. Behrends, V. et al. 2014. A metabolic trade-off between phosphate and glucose utilization in *Escherichia coli. Molecular bioSystems* 10, 2820-2822.

33. Beisel, C. L. & Afroz, T. 2016. Rethinking the hierarchy of sugar utilization in bacteria. *Journal of Bacteriology* 198, 374-376.

34. Bekker, M. et al. 2009. Respiration of *Escherichia coli* can be fully uncoupled via the nonelectrogenic terminal cytochrome bd-II oxidase. *Journal of Bacteriology* 191, 5510-5517.

35. Benarroch, J. M. & Asally, M. 2020. The microbiologist's guide to membrane potential dynamics. *Trends in Microbiology* 28, 304-314.

36. Bennett, J. H. 1983. *Natural Selection, Heredity, and Eugenics, Including Selected Correspondence of R. A. Fisher with Leonard Darwin and Others.* Oxford: Oxford University Press.

37. Bergeron-Sandoval, L.-P., Safaee, N. & Michnick, S. W. 2016. Mechanisms and consequences of macromolecular phase separation. *Cell* 165, 1067-1079.

38. Bergk Pinto, B. et al. 2019. Do organic substrates drive microbial community interactions in Arctic snow? *Frontiers in Microbiology* 10, 2492.

39. Berney, M. et al. 2006. Specific growth rate determines the sensitivity of *Escherichia coli* to thermal, UVA, and solar disinfection. *Applied and Environmental Microbiology* 72, 2586-2593.

40. Bernstein, W. J. & Wilkinson, D. 1997. Diversification, rebalancing, and the geometric mean frontier. *Social Science Research Network,* doi: 10.2139/ssrn.53503.

41. Bjerg, J. T. et al. 2016. Motility of electric cable bacteria. *Applied and Environmental Microbiology* 82, 3816-3821.

42. Bleier, L. & Dröse, S. 2013. Superoxide generation by complex III: from mechanistic rationales to functional consequences. *Biochimica et Biophysica Acta (BBA)-Bioenergetics* 1827, 1320-1331.

43. Boltzmann, L. 1974. *Theoretical Physics and Philosophical Problems: Selected Writings*. Ed. by B. F. McGuinness. Vol. 5. Vienna Circle Collection. Dordrecht, Netherlands: Springer.

44. Borguet, S. & Leonard, O. 2008. The Fisher information matrix as a relevant tool for sensor selection in engine health monitoring. *International Journal of Rotating Machinery* 2008, 784749.

45. Boschker, H. T. S. et al. 2021. Efficient long-range conduction in cable bacteria through nickel protein wires. *Nature Communications* 12, 3996.

46. Brady, K. U., Kruckeberg, A. R. & Bradshaw Jr, H. D. 2005. Evolutionary ecology of plant adaptation to serpentine soils. *Annual Review of Ecology and Systematics* 36, 243-266.

47. Branscomb, E., Biancalani, T., Goldenfeld, N. & Russell, M. 2017. Escapement mechanisms and the conversion of disequilibria; the engines of creation. *Physics Reports* 677, 1-60.

48. Bräsen, C., Esser, D., Rauch, B. & Siebers, B. 2014. Carbohydrate metabolism in Archaea: current insights into unusual enzymes and pathways and their regulation. *Microbiology and Molecular Biology Reviews* 78, 89-175.

49. Brauer, M. J., Saldanha, A. J., Dolinski, K. & Botstein, D. 2005. Homeostatic adjustment and metabolic remodeling in glucose-limited yeast cultures. *Molecular Biology of the Cell* 16, 2503-2517.

50. Brinkmeyer, R. et al. 2003. Diversity and structure of bacterial communities in Arctic versus Antarctic pack ice. *Applied and Environmental Microbiology* 69, 6610-6619.

51. Bruggeman, F. J., Planqué, R., Molenaar, D. & Teusink, B. 2020. Searching for principles of microbial physiology. *FEMS Microbiology Reviews* 44, 821-844.

52. Bull, J. J. 1994. Perspective: virulence. *Evolution* 48, 1423-1437.

53. Burdorf, L. D. et al. 2017. Long-distance electron transport occurs globally in marine sediments. *Biogeosciences* 14, 683-701.

54. Burt, A. & Trivers, R. 2008. *Genes in Conflict: The Biology of Selfish Genetic Elements.* Cambridge, MA: Belknap Press.

55. Butaitė, E., Baumgartner, M., Wyder, S. & Kümmerli, R. 2017. Siderophore cheating and cheating resistance shape competition for iron in soil and freshwater *Pseudomonas* communities. *Nature Communications* 8, 1-12.

56. Cain, J. A., Solis, N. & Cordwell, S. J. 2014. Beyond gene expression: the impact of protein post-translational modifications in bacteria. *Journal of Proteomics* 97, 265-286.

57. Carmel-Harel, O. & Storz, G. 2000. Roles of the glutathione- and thioredoxin-dependent reduction systems in the *Escherichia coli* and *Saccharomyces cerevisiae* responses to oxidative stress. *Annual Reviews in Microbiology* 54, 439-461.

58. Carter Jr, C. W. 2020. Escapement mechanisms: efficient free energy transduction by reciprocally-coupled gating. *Proteins: Structure, Function, and Bioinformatics* 88, 710-717.

59. Carter Jr, C. W. & Wills, P. R. 2021. Reciprocally-coupled gating: strange loops in bioenergetics, genetics, and catalysis. *Biomolecules* 11, 265.

60. Chakraborty, R., V. Braun, K. Hantke & P. Cornelis, eds. 2013. *Iron Uptake in Bacteria with Emphasis on E. coli and Pseudomonas.* New York: Springer Science & Business Media.

61. Charlesworth, B. 1994. *Evolution in Age-Structured Populations.* 2nd ed. Cambridge: Cambridge University Press.

62. Charlesworth, B., Lande, R. & Slatkin, M. 1982. A neo-Darwinian commentary on macroevolution. *Evolution* 36, 474-498.

63. Chavarría, M., Nikel, P. I., Pérez-Pantoja, D. & Lorenzo, V. de. 2013. The Entner–Doudoroff pathway empowers *Pseudomonas putida* KT2440 with a high tolerance to oxidative stress. *Environmental Microbiology* 15, 1772-1785.

64. Chen, X. et al. 2016. The Entner–Doudoroff pathway is an overlooked glycolytic route in cyanobacteria and plants. *Proceedings of the National Academy of Sciences* 113, 5441-5446.

65. Cheng, C. et al. 2019. Laboratory evolution reveals a two-dimensional rate-yield tradeoff in microbial metabolism. *PLoS Computational Biology* 15, e1007066.

66. Chubukov, V., Gerosa, L., Kochanowski, K. & Sauer, U. 2014. Coordination of microbial metabolism. *Nature Reviews Microbiology* 12, 327–340.

67. Conrad, M. et al. 2014. Nutrient sensing and signaling in the yeast *Saccharomyces cerevisiae*. *FEMS Microbiology Reviews* 38, 254–299.

68. Cornelis, P. & Bodilis, J. 2009. A survey of TonB-dependent receptors in fluorescent pseudomonads. *Environmental Microbiology Reports* 1, 256–262.

69. Cornelis, P. & Matthijs, S. 2002. Diversity of siderophore-mediated iron uptake systems in fluorescent pseudomonads: not only pyoverdines. *Environmental Microbiology* 4, 787–798.

70. Cosentino, C. & Bates, D. 2011. *Feedback Control in Systems Biology*. Boca Raton, Florida: CRC Press.

71. Crow, J. F. & Kimura, M. 1970. *An Introduction to Population Genetics Theory*. Minneapolis, MN: Burgess.

72. Curtin, K., Duerre, S., Fitzpatrick, B. & Meyer, P. 2011. *Biological Nutrient Removal*. St. Paul, MN: Minnesota Pollution Control.

73. D'Souza, G. et al. 2018. Ecology and evolution of metabolic cross-feeding interactions in bacteria. *Natural Product Reports* 35, 455–488.

74. Darwin, C. 1860. *On the Origin of Species by Means of Natural Selection*. 2nd ed. London: John Murray.

75. Darwin, C. 1877. *The Various Contrivances by which Orchids are Fertilised by Insects*. Chicago: University of Chicago Press.

76. Darwin, F. 1903. *More Letters of Charles Darwin*. Vol. 1. London: Murray.

77. Das, K. & Roychoudhury, A. 2014. Reactive oxygen species (ROS) and response of antioxidants as ROS-scavengers during environmental stress in plants. *Frontiers in Environmental Science* 2, 53.

78. De Groot, D. H. et al. 2020. The common message of constraint-based optimization approaches: overflow metabolism is caused by two growth-limiting constraints. *Cellular and Molecular Life Sciences* 77, 441–453.

79. De Paepe, M. & Taddei, F. 2006. Viruses' life history: towards a mechanistic basis of a trade-off between survival and reproduction among phages. *PLoS Biology* 4, e193.

80. Dempster, E. R. 1955. Maintenance of genetic heterogeneity. *Cold Spring Harbor Symposia on Quantitative Biology* 20, 25–32.

81. Deutscher, J., Francke, C. & Postma, P. W. 2006. How phosphotransferase system-related protein phosphorylation regulates carbohydrate metabolism in bacteria. *Microbiology and Molecular Biology Reviews* 70, 939–1031.

82. DeWitt, T. J. & Scheiner, S. M. 2004. *Phenotypic Plasticity: Functional and Conceptual Approaches*. New York: Oxford University Press.

83. Di Bartolomeo, F. et al. 2020. Absolute yeast mitochondrial proteome quantification reveals trade-off between biosynthesis and energy generation during diauxic shift. *Proceedings of the National Academy of Sciences* 117, 7524–7535.

84. Diard, M. et al. 2013. Stabilization of cooperative virulence by the expression of an avirulent phenotype. *Nature* 494, 353–356.

85. Dickinson, E. 1955. *The Poems of Emily Dickinson*. Ed. by T. H. Johnson. Cambridge, MA: Harvard University Press.

86. Dill, K. & Bromberg, S. 2012. *Molecular Driving Forces: Statistical Thermodynamics in Biology, Chemistry, Physics, and Nanoscience*. New York: Garland Science.

87. Doebeli, M. & Knowlton, N. 1998. The evolution of interspecific mutualisms. *Proceedings of the National Academy of Sciences* 95, 8676–8680.

88. Dorofeev, A., Nikolaev, Y. A., Mardanov, A. & Pimenov, N. 2020. Role of phosphate-accumulating bacteria in biological phosphorus removal from wastewater. *Applied Biochemistry and Microbiology* 56, 1–14.

89. Doyle, J. C., Francis, B. A. & Tannenbaum, A. R. 2009. *Feedback Control Theory*. Mineola, NY: Dover Publications.

90. Duar, R. M. et al. 2017. Lifestyles in transition: evolution and natural history of the genus *Lactobacillus*. *FEMS Microbiology Reviews* 41, S27–S48.

91. Dunn, J. C. et al. 2015. Evolutionary trade-off between vocal tract and testes dimensions in howler monkeys. *Current Biology* 25, 2839–2844.

92. Ebrahimi, A., Schwartzman, J. & Cordero, O. X. 2019. Cooperation and spatial self-organization determine rate and efficiency of particulate organic matter degradation in marine bacteria. *Proceedings of the National Academy of Sciences* 116, 23309–23316.

93. Elena, S. F. & Lenski, R. E. 2003. Evolution experiments with microorganisms: the dynamics and genetic bases of adaptation. *Nature Review Genetics* 4, 457-469.

94. Elowitz, M. B., Levine, A. J., Siggia, E. D. & Swain, P. S. 2002. Stochastic gene expression in a single cell. *Science* 297, 1183-1186.

95. Elser, J. et al. 2003. Growth rate-stoichiometry couplings in diverse biota. *Ecology Letters* 6, 936-943.

96. Enjalbert, B. et al. 2017. Acetate fluxes in *Escherichia coli* are determined by the thermodynamic control of the Pta-AckA pathway. *Scientific Reports* 7, 42135.

97. Erickson, D. W. et al. 2017. A global resource allocation strategy governs growth transition kinetics of *Escherichia coli. Nature* 551, 119-123.

98. Escalante-Chong, R. et al. 2015. Galactose metabolic genes in yeast respond to a ratio of galactose and glucose. *Proceedings of the National Academy of Sciences* 112, 1636-1641.

99. Ewens, W. J. 2010. *Mathematical Population Genetics: I. Theoretical Introduction.* 2nd ed. New York: Springer-Verlag.

100. Fell, D. 1997. *Understanding the Control of Metabolism.* London: Portland Press.

101. Ferenci, T. 2001. Hungry bacteria—definition and properties of a nutritional state. *Environmental Microbiology* 3, 605-611.

102. Ferenci, T. 2005. Maintaining a healthy SPANC balance through regulatory and mutational adaptation. *Molecular Microbiology* 57, 1-8.

103. Ferenci, T. 2007. Bacterial physiology, regulation and mutational adaptation in a chemostat environment. *Advances in Microbial Physiology* 53, 169-229.

104. Ferenci, T. 2016. Trade-off mechanisms shaping the diversity of bacteria. *Trends in Microbiology* 24, 209-223.

105. Finster, K. 2008. Microbiological disproportionation of inorganic sulfur compounds. *Journal of Sulfur Chemistry* 29, 281-292.

106. Fisher, R. A. 1958. *The Genetical Theory of Natural Selection.* 2nd ed. New York: Dover Publications.

107. Fisher, R. A. 1971. *The Design of Experiments.* 9th ed. New York: Macmillan.

108. Flamholz, A. et al. 2013. Glycolytic strategy as a tradeoff between energy yield and protein cost. *Proceedings of the National Academy of Sciences* 110, 10039-10044.

109. Flechsler, J. et al. 2021. Functional compartmentalization and metabolic separation in a prokaryotic cell. *Proceedings of the National Academy of Sciences* 118, e2022114118.

110. Foster, K. R. & Wenseleers, T. 2006. A general model for the evolution of mutualisms. *Journal of Evolutionary Biology* 19, 1283-1293.

111. Frank, S. A. 1986. Dispersal polymorphisms in subdivided populations. *Journal of Theoretical Biology* 122, 303-309.

112. Frank, S. A. 1986. Hierarchical selection theory and sex ratios I. General solutions for structured populations. *Theoretical Population Biology* 29, 312-342.

113. Frank, S. A. 1994. Genetics of mutualism: the evolution of altruism between species. *Journal of Theoretical Biology* 170, 393-400.

114. Frank, S. A. 1994. Kin selection and virulence in the evolution of protocells and parasites. *Proceedings of the Royal Society of London B* 258, 153-161.

115. Frank, S. A. 1995. Mutual policing and repression of competition in the evolution of cooperative groups. *Nature* 377, 520-522.

116. Frank, S. A. 1995. The origin of synergistic symbiosis. *Journal of Theoretical Biology* 176, 403-410.

117. Frank, S. A. 1996. Models of parasite virulence. *Quarterly Review of Biology* 71, 37-78.

118. Frank, S. A. 1996. Policing and group cohesion when resources vary. *Animal Behaviour* 52, 1163-1169.

119. Frank, S. A. 1997. Models of symbiosis. *American Naturalist* 150, S80-S99.

120. Frank, S. A. 1997. Multivariate analysis of correlated selection and kin selection, with an ESS maximization method. *Journal of Theoretical Biology* 189, 307-316.

121. Frank, S. A. 1997. The Price equation, Fisher's fundamental theorem, kin selection, and causal analysis. *Evolution* 51, 1712-1729.

122. Frank, S. A. 1998. *Foundations of Social Evolution*. Princeton, NJ: Princeton University Press.

123. Frank, S. A. 2003. Repression of competition and the evolution of cooperation. *Evolution* 57, 693-705.

124. Frank, S. A. 2004. Genetic variation in cancer predisposition: mutational decay of a robust genetic control network. *Proceedings of the National Academy of Sciences* 101, 8061-8065.

125. Frank, S. A. 2007. Maladaptation and the paradox of robustness in evolution. *PLoS ONE* 2, e1021.

126. Frank, S. A. 2009. Evolutionary foundations of cooperation and group cohesion. In: *Games, Groups, and the Global Good.* Ed. by S. A. Levin. New York: Springer, 3-40.

127. Frank, S. A. 2010. A general model of the public goods dilemma. *Journal of Evolutionary Biology* 23, 1245-1250.

128. Frank, S. A. 2010. Demography and the tragedy of the commons. *Journal of Evolutionary Biology* 23, 32-39.

129. Frank, S. A. 2010. Microbial secretor-cheater dynamics. *Philosophical Transactions of the Royal Society B* 365, 2515-2522.

130. Frank, S. A. 2010. The trade-off between rate and yield in the design of microbial metabolism. *Journal of Evolutionary Biology* 23, 609-613.

131. Frank, S. A. 2011. Natural selection. I. Variable environments and uncertain returns on investment. *Journal of Evolutionary Biology* 24, 2299-2309.

132. Frank, S. A. 2011. Natural selection. II. Developmental variability and evolutionary rate. *Journal of Evolutionary Biology* 24, 2310-2320.

133. Frank, S. A. 2012. Natural selection. III. Selection versus transmission and the levels of selection. *Journal of Evolutionary Biology* 25, 227-243.

134. Frank, S. A. 2013. Evolution of robustness and cellular stochasticity of gene expression. *PLoS Biology* 11, e1001578.

135. Frank, S. A. 2013. Microbial evolution: regulatory design prevents cancer-like overgrowths. *Current Biology* 23, R343-R346.

136. Frank, S. A. 2013. Natural selection. VII. History and interpretation of kin selection theory. *Journal of Evolutionary Biology* 26, 1151-1184.

137. Frank, S. A. 2017. Receptor uptake arrays for vitamin B12, siderophores, and glycans shape bacterial communities. *Ecology and Evolution* 7, 10175-10195.

138. Frank, S. A. 2018. *Control Theory Tutorial: Basic Concepts Illustrated by Software Examples.* Cham, Switzerland: Springer.

139. Frank, S. A. 2019. Evolutionary design of regulatory control. I. A robust control theory analysis of tradeoffs. *Journal of Theoretical Biology* 463, 121–137.

140. Frank, S. A. 2019. Evolutionary design of regulatory control. II. Robust error-correcting feedback increases genetic and phenotypic variability. *Journal of Theoretical Biology* 468, 72–81.

141. Frank, S. A. 2020. Metabolic heat in microbial conflict and cooperation. *Frontiers in Ecology and Evolution* 8, 275.

142. Frank, S. A. & Rosner, M. R. 2012. Nonheritable cellular variability accelerates the evolutionary processes of cancer. *PLoS Biology* 10, e1001296.

143. Frank, S. A. & Schmid-Hempel, P. 2008. Mechanisms of pathogenesis and the evolution of parasite virulence. *Journal of Evolutionary Biology* 21, 396–404.

144. Frank, S. A. & Slatkin, M. 1990. Evolution in a variable environment. *American Naturalist* 136, 244–260.

145. Frumkin, I. et al. 2017. Gene architectures that minimize cost of gene expression. *Molecular Cell* 65, 142–153.

146. Gao, L. et al. 2020. Diverse enzymatic activities mediate antiviral immunity in prokaryotes. *Science* 369, 1077–1084.

147. Garcia, N. S. et al. 2018. High variability in cellular stoichiometry of carbon, nitrogen, and phosphorus within classes of marine eukaryotic phytoplankton under sufficient nutrient conditions. *Frontiers in Microbiology* 9, 543.

148. García-Bayona, L. & Comstock, L. E. 2018. Bacterial antagonism in host-associated microbial communities. *Science* 361, eaat2456.

149. Garcia-Garcera, M. & Rocha, E. P. 2020. Community diversity and habitat structure shape the repertoire of extracellular proteins in bacteria. *Nature Communications* 11, 1–11.

150. Gardner, A. & West, S. 2006. Demography, altruism, and the benefits of budding. *Journal of Evolutionary Biology* 19, 1707–1716.

151. Geerlings, N. M. et al. 2020. Division of labor and growth during electrical cooperation in multicellular cable bacteria. *Proceedings of the National Academy of Sciences* 117, 5478–5485.

152. Geerlings, N. M. et al. 2021. Cell cycle, filament growth and synchronized cell division in multicellular cable bacteria. *Frontiers in Microbiology* 12, 620807.

153. Gillespie, J. H. 1977. Natural selection for variances in offspring numbers: a new evolutionary principle. *American Naturalist* 111, 1010-1014.

154. Gillespie, J. H. 1994. *The Causes of Molecular Evolution.* New York: Oxford University Press.

155. Glasser, N. R., Kern, S. E. & Newman, D. K. 2014. Phenazine redox cycling enhances anaerobic survival in *Pseudomonas aeruginosa* by facilitating generation of ATP and a proton-motive force. *Molecular Microbiology* 92, 399-412.

156. Goelzer, A. & Fromion, V. 2011. Bacterial growth rate reflects a bottleneck in resource allocation. *Biochimica et Biophysica Acta (BBA)-General Subjects* 1810, 978-988.

157. Granato, E. T., Meiller-Legrand, T. A. & Foster, K. R. 2019. The evolution and ecology of bacterial warfare. *Current Biology* 29, R521-R537.

158. Grangeasse, C., Nessler, S. & Mijakovic, I. 2012. Bacterial tyrosine kinases: evolution, biological function and structural insights. *Philosophical Transactions of the Royal Society B* 367, 2640-2655.

159. Grimbergen, A. J., Siebring, J., Solopova, A. & Kuipers, O. P. 2015. Microbial bet-hedging: the power of being different. *Current Opinion in Microbiology* 25, 67-72.

160. Grondin, J. M. et al. 2017. Polysaccharide utilization loci: fuelling microbial communities. *Journal of Bacteriology* 199, e00860-16.

161. Großkopf, T. & Soyer, O. S. 2016. Microbial diversity arising from thermodynamic constraints. *ISME Journal* 10, 2725-2733.

162. Großkopf, T. et al. 2016. A stable genetic polymorphism underpinning microbial syntrophy. *ISME Journal* 10, 2844-2853.

163. Grüning, N.-M. et al. 2011. Pyruvate kinase triggers a metabolic feedback loop that controls redox metabolism in respiring cells. *Cell Metabolism* 14, 415-427.

164. Gude, S. et al. 2020. Bacterial coexistence driven by motility and spatial competition. *Nature* 578, 588-592.

165. Gyorgy, A. et al. 2015. Isocost lines describe the cellular economy of genetic circuits. *Biophysical Journal* 109, 639-646.

166. Hamilton, W. D. 1964. The genetical evolution of social behaviour. I. *Journal of Theoretical Biology* 7, 1-16.

167. Hamilton, W. D. 1970. Selfish and spiteful behaviour in an evolutionary model. *Nature* 228, 1218-1220.

168. Hamilton, W. D. 1975. Innate social aptitudes of man: an approach from evolutionary genetics. In: *Biosocial Anthropology*. Ed. by R. Fox. New York: Wiley, 133–155.

169. Hamilton, W. D. & May, R. M. 1977. Dispersal in stable habitats. *Nature* 269, 578–581.

170. Harcombe, W. 2010. Novel cooperation experimentally evolved between species. *Evolution* 64, 2166–2172.

171. Harcombe, W. R. et al. 2018. Evolution of bidirectional costly mutualism from byproduct consumption. *Proceedings of the National Academy of Sciences* 115, 12000–12004.

172. Hardin, G. 1968. The tragedy of the commons. *Science* 162, 1243–1248.

173. Harvey, P. H. & Pagel, M. D. 1991. *The Comparative Method in Evolutionary Biology*. Oxford: Oxford University Press.

174. Hausser, J., Mayo, A., Keren, L. & Alon, U. 2019. Central dogma rates and the trade-off between precision and economy in gene expression. *Nature Communications* 10, 1–15.

175. Hector, T. E. & Booksmythe, I. 2019. Digest: Little evidence exists for a virulence-transmission trade-off. *Evolution* 73, 858–859.

176. Hedrick, P. W. 2006. Genetic polymorphism in heterogeneous environments: the age of genomics. *Annual Review of Ecology and Systematics* 37, 67–93.

177. Heijden, J. van der et al. 2016. *Salmonella* rapidly regulates membrane permeability to survive oxidative stress. *mBio* 7, e01238-16.

178. Hibbing, M. E., Fuqua, C., Parsek, M. R. & Peterson, S. B. 2010. Bacterial competition: surviving and thriving in the microbial jungle. *Nature Reviews Microbiology* 8, 15–25.

179. Hider, R. C. & Kong, X. 2010. Chemistry and biology of siderophores. *Natural Product Reports* 27, 637–657.

180. Hijum, S. A. van, Medema, M. H. & Kuipers, O. P. 2009. Mechanisms and evolution of control logic in prokaryotic transcriptional regulation. *Microbiology and Molecular Biology Reviews* 73, 481–509.

181. Hill, T. L. 1983. Some general principles in free energy transduction. *Proceedings of the National Academy of Sciences* 80, 2922–2925.

182. Hill, T. L. 1989. *Free Energy Transduction and Biochemical Cycle Kinetics*. New York: Springer-Verlag.

183. Hinton, G. E. & Nowlan, S. J. 1987. How learning can guide evolution. *Complex Systems* 1, 495–502.

184. Holmes, D. E. et al. 2019. A membrane-bound cytochrome enables *Methanosarcina acetivorans* to conserve energy from extracellular electron transfer. *mBio* 10, e00789-19.

185. Hoshino, T. et al. 2020. Global diversity of microbial communities in marine sediment. *Proceedings of the National Academy of Sciences* 117, 27587–27597.

186. Hou, J., Scalcinati, G., Oldiges, M. & Vemuri, G. N. 2010. Metabolic impact of increased NADH availability in *Saccharomyces cerevisiae*. *Applied and Environmental. Microbiology* 76, 851–859.

187. Hühner, J. et al. 2015. Quantification of riboflavin, flavin mononucleotide, and flavin adenine dinucleotide in mammalian model cells by CE with LED-induced fluorescence detection. *Electrophoresis* 36, 518–525.

188. Hui, S. et al. 2015. Quantitative proteomic analysis reveals a simple strategy of global resource allocation in bacteria. *Molecular Systems Biology* 11, 784.

189. Hunter, T. 1995. Protein kinases and phosphatases: the yin and yang of protein phosphorylation and signaling. *Cell* 80, 225–236.

190. Hwa, T. & Sauer, U. 2018. Editorial overview: Current Opinion in Microbiology 2018 special issue 'Microbial systems biology, vol. 45'. *Current Opinion in Microbiology* 45, vi–viii.

191. Ihssen, J. & Egli, T. 2005. Global physiological analysis of carbon- and energy-limited growing *Escherichia coli* confirms a high degree of catabolic flexibility and preparedness for mixed substrate utilization. *Environmental Microbiology* 7, 1568–1581.

192. Imlay, J. A. 2019. Where in the world do bacteria experience oxidative stress? *Environmental Microbiology* 21, 521–530.

193. Jacobson, T. B. et al. 2020. In vivo thermodynamic analysis of glycolysis in *Clostridium thermocellum* and *Thermoanaerobacterium saccharolyticum* using ^{13}C and ^{2}H tracers. *mSystems* 5, e00736-19.

194. Janssens, G. E. & Veenhoff, L. M. 2016. The natural variation in lifespans of single yeast cells is related to variation in cell size, ribosomal protein, and division time. *PloS ONE* 11, e0167394.

195. Jeuken, L. J., Hards, K. & Nakatani, Y. 2020. Extracellular electron transfer: respiratory or nutrient homeostasis? *Journal of Bacteriology* 202, e00029-20.

196. Johnson, K. J. & Rose-Pehrsson, S. L. 2015. Sensor array design for complex sensing tasks. *Annual Review of Analytical Chemistry* 8, 287-310.

197. Jørgensen, B. B., Findlay, A. J. & Pellerin, A. 2019. The biogeochemical sulfur cycle of marine sediments. *Frontiers in Microbiology* 10, 849.

198. Kaczmarczyk, A. et al. 2020. Precise timing of transcription by c-di-GMP coordinates cell cycle and morphogenesis in *Caulobacter. Nature Communications* 11, 816.

199. Kaern, M., Elston, T. C., Blake, W. J. & Collins, J. J. 2005. Stochasticity in gene expression: from theories to phenotypes. *Nature Reviews Genetics* 6, 451-464.

200. Kafri, M., Metzl-Raz, E., Jona, G. & Barkai, N. 2016. The cost of protein production. *Cell Reports* 14, 22-31.

201. Kamp, F., Welch, G. R. & Westerhoff, H. V. 1988. Energy coupling and Hill cycles in enzymatic processes. *Cell Biophysics* 12, 201-236.

202. Kamrad, S. et al. 2020. Pyruvate kinase variant of fission yeast tunes carbon metabolism, cell regulation, growth and stress resistance. *Molecular Systems Biology* 16, e9270.

203. Kappler, A. et al. 2021. An evolving view on biogeochemical cycling of iron. *Nature Reviews Microbiology*, 1-15.

204. Kassen, R. & Rainey, P. B. 2004. The ecology and genetics of microbial diversity. *Annual Review of Microbiology* 58, 207-231.

205. Keller, L., ed. 1999. *Levels of Selection in Evolution.* Princeton, NJ: Princeton University Press.

206. Kellner, S. et al. 2018. Genome size evolution in the Archaea. *Emerging Topics in Life Sciences* 2, 595-605.

207. Ketola, T., Mikonranta, L. & Mappes, J. 2016. Evolution of bacterial life-history traits is sensitive to community structure. *Evolution* 70, 1334-1341.

208. Kim, B. H. & Gadd, G. M. 2019. *Prokaryotic Metabolism and Physiology.* Cambridge: Cambridge University Press.

209. Kim, J. et al. 2020. Trade-offs between gene expression, growth and phenotypic diversity in microbial populations. *Current Opinion in Biotechnology* 62, 29-37.

210. Kjeldsen, K. U. et al. 2019. On the evolution and physiology of cable bacteria. *Proceedings of the National Academy of Sciences* 116, 19116-19125.

211. Klausmeier, C. A., Litchman, E., Daufresne, T. & Levin, S. A. 2004. Optimal nitrogen-to-phosphorus stoichiometry of phytoplankton. *Nature* 429, 171-174.

212. Koirala, S., Wang, X. & Rao, C. V. 2016. Reciprocal regulation of l-arabinose and d-xylose metabolism in *Escherichia coli. Journal of Bacteriology* 198, 386-393.

213. Kotloski, N. J. & Gralnick, J. A. 2013. Flavin electron shuttles dominate extracellular electron transfer by *Shewanella oneidensis. mBio* 4, e00553-12.

214. Kouzuma, A., Kato, S. & Watanabe, K. 2015. Microbial interspecies interactions: recent findings in syntrophic consortia. *Frontiers in Microbiology* 6, 477.

215. Kozuch, S. 2015. Steady state kinetics of any catalytic network: graph theory, the energy span model, the analogy between catalysis and electrical circuits, and the meaning of "mechanism". *ACS Catalysis* 5, 5242-5255.

216. Kramer, J., Özkaya, Ö. & Kümmerli, R. 2020. Bacterial siderophores in community and host interactions. *Nature Reviews Microbiology* 18, 152-163.

217. Kumar, A. et al. 2017. The ins and outs of microorganism-electrode electron transfer reactions. *Nature Reviews Chemistry* 1, 1-13.

218. Kümmerli, R. et al. 2014. Habitat structure and the evolution of diffusible siderophores in bacteria. *Ecology Letters* 17, 1536-1544.

219. LaCroix, R. A. et al. 2015. Use of adaptive laboratory evolution to discover key mutations enabling rapid growth of *Escherichia coli* K-12 MG1655 on glucose minimal medium. *Applied and Environmental Microbiology* 81, 17-30.

220. Lakatos, I. 1978. *The Methodology of Scientific Research Programs: Philosophical Papers.* Ed. by J. Worrall & G. Currie. Vol. 1. Cambridge: Cambridge University Press.

221. Lanciano, P. et al. 2013. Molecular mechanisms of superoxide production by complex III: a bacterial versus human mitochondrial comparative case study. *Biochimica et Biophysica Acta (BBA)-Bioenergetics* 1827, 1332-1339.

222. Lanczos, C. 1986. *The Variational Principles of Mechanics.* 4th ed. New York: Dover Publications.

223. Lande, R. 2008. Adaptive topography of fluctuating selection in a Mendelian population. *Journal of Evolutionary Biology* 21, 1096-1105.

224. Lange, R. & Hengge-Aronis, R. 1991. Growth phase-regulated expression of bolA and morphology of stationary-phase *Escherichia coli* cells are controlled by the novel sigma factor sigma S. *Journal of Bacteriology* 173, 4474-4481.

225. Łapińska, U. et al. 2019. Bacterial ageing in the absence of external stressors. *Philosophical Transactions of the Royal Society B* 374, 20180442.

226. Lawrence, D. et al. 2012. Species interactions alter evolutionary responses to a novel environment. *PLoS Biology* 10, e1001330.

227. Lazebnik, Y. 2002. Can a biologist fix a radio?—Or, what I learned while studying apoptosis. *Cancer Cell* 2, 179-182.

228. Lee, H.-S. et al. 2016. The roles of biofilm conductivity and donor substrate kinetics in a mixed-culture biofilm anode. *Environmental Science & Technology* 50, 12799-12807.

229. Lennon, J. T. & Jones, S. E. 2011. Microbial seed banks: the ecological and evolutionary implications of dormancy. *Nature Reviews Microbiology* 9, 119-130.

230. Lenski, R. E. & May, R. M. 1994. The evolution of virulence in parasites and pathogens: reconciliation between two competing hypotheses. *Journal Theoretical Biology* 169, 253-265.

231. Lenski, R. E. & Travisano, M. 1994. Dynamics of adaptation and diversification: a 10,000-generation experiment with bacterial populations. *Proceedings of the National Academy of Sciences* 91, 6808-6814.

232. Leonard, C. J., Aravind, L. & Koonin, E. V. 1998. Novel families of putative protein kinases in bacteria and archaea: evolution of the "eukaryotic" protein kinase superfamily. *Genome Research* 8, 1038-1047.

233. Leslie, M. 2021. Separation anxiety. *Science* 371, 336-338.

234. Levene, H. 1953. Genetic equilibrium when more than one ecological niche is available. *American Naturalist* 87, 331–333.

235. Leventhal, G. E., Ackermann, M. & Schiessl, K. T. 2019. Why microbes secrete molecules to modify their environment: the case of iron-chelating siderophores. *Journal of the Royal Society Interface* 16, 20180674.

236. Levin, B. R. & Bull, J. J. 1994. Short-sighted evolution and the virulence of pathogenic microorganisms. *Trends in Microbiology* 2, 76–81.

237. Levin, P. A. & Angert, E. R. 2015. Small but mighty: cell size and bacteria. *Cold Spring Harbor Perspectives in Biology* 7, a019216.

238. Levy, E. D., De, S. & Teichmann, S. A. 2012. Cellular crowding imposes global constraints on the chemistry and evolution of proteomes. *Proceedings of the National Academy of Sciences* 109, 20461–20466.

239. Lewis, N. E. et al. 2010. Omic data from evolved *E. coli* are consistent with computed optimal growth from genome-scale models. *Molecular Systems Biology* 6, 390.

240. Li, G.-W., Burkhardt, D., Gross, C. & Weissman, J. S. 2014. Quantifying absolute protein synthesis rates reveals principles underlying allocation of cellular resources. *Cell* 157, 624–635.

241. Li, Y. et al. 2020. A programmable fate decision landscape underlies single-cell aging in yeast. *Science* 369, 325–329.

242. Light, S. H. et al. 2018. A flavin-based extracellular electron transfer mechanism in diverse Gram-positive bacteria. *Nature* 562, 140–144.

243. Light, S. H. et al. 2019. Extracellular electron transfer powers flavinylated extracellular reductases in Gram-positive bacteria. *Proceedings of the National Academy of Sciences* 116, 26892–26899.

244. Lindquist, S. & Craig, E. A. 1988. The heat-shock proteins. *Annual Review of Genetics* 22, 631–677.

245. Linzner, K. A., Kent, A. G. & Martiny, A. C. 2018. Evolutionary pathway determines the stoichiometric response of *Escherichia coli* adapted to high temperature. *Frontiers in Ecology and Evolution* 5, 173.

246. Lipson, D. A. 2015. The complex relationship between microbial growth rate and yield and its implications for ecosystem processes. *Frontiers in Microbiology* 6, 615.

247. Litchman, E., Edwards, K. F. & Klausmeier, C. A. 2015. Microbial resource utilization traits and trade-offs: implications for community structure, functioning, and biogeochemical impacts at present and in the future. *Frontiers in Microbiology* 6, 254.

248. Liu, F. et al. 2021. Cable bacteria extend the impacts of elevated dissolved oxygen into anoxic sediments. *ISME Journal* 15, 1551-1563.

249. Löffler, M. et al. 2017. Switching between nitrogen and glucose limitation: unraveling transcriptional dynamics in *Escherichia coli*. *Journal of Biotechnology* 258, 2-12.

250. Logan, B. E., Rossi, R., Saikaly, P. E., et al. 2019. Electroactive microorganisms in bioelectrochemical systems. *Nature Reviews Microbiology* 17, 307-319.

251. Lovley, D. R. 2017. Happy together: microbial communities that hook up to swap electrons. *ISME Journal* 11, 327-336.

252. Lovley, D. R. 2017. Syntrophy goes electric: direct interspecies electron transfer. *Annual Review of Microbiology* 71, 643-664.

253. Lovley, D. R. & Holmes, D. E. 2020. Protein nanowires: the electrification of the microbial world and maybe our own. *Journal of Bacteriology* 202, e00331-20.

254. Lynch, M. 2007. *The Origins of Genome Architecture*. Sunderland, MA: Sinauer Associates.

255. Lyon, B. E. & Montgomerie, R. 2012. Sexual selection is a form of social selection. *Philosophical Transactions of the Royal Society B* 367, 2266-2273.

256. Machelart, A. et al. 2020. Convergent evolution of zoonotic *Brucella* species toward the selective use of the pentose phosphate pathway. *Proceedings of the National Academy of Sciences* 117, 26374-26381.

257. MacLean, R. C. 2008. The tragedy of the commons in microbial populations: insights from theoretical, comparative and experimental studies. *Heredity* 100, 471-477.

258. MacLean, R. C. & Gudelj, I. 2006. Resource competition and social conflict in experimental populations of yeast. *Nature* 441, 498-501.

259. Malkin, S. Y. et al. 2014. Natural occurrence of microbial sulphur oxidation by long-range electron transport in the seafloor. *ISME Journal* 8, 1843-1854.

260. Mallory, J. D., Kolomeisky, A. B. & Igoshin, O. A. 2020. Kinetic control of stationary flux ratios for a wide range of biochemical processes. *Proceedings of the National Academy of Sciences* 117, 8884-8889.

261. Malvankar, N. S. et al. 2012. Electrical conductivity in a mixed-species biofilm. *Applied and Environmental Microbiology* 78, 5967-5971.

262. Marshall, A. 1966. *Principles of Economics.* London: Macmillan.

263. Martens, E. C. et al. 2011. Recognition and degradation of plant cell wall polysaccharides by two human gut symbionts. *PLOS Biology* 9, 1-16.

264. Martino, M. E. et al. 2016. Nomadic lifestyle of *Lactobacillus plantarum* revealed by comparative genomics of 54 strains isolated from different habitats. *Environmental Microbiology* 18, 4974-4989.

265. Martins, G., Salvador, A. F., Pereira, L. & Alves, M. M. 2018. Methane production and conductive materials: a critical review. *Environmental Science & Technology* 52, 10241-10253.

266. Mauri, M. & Klumpp, S. 2014. A model for sigma factor competition in bacterial cells. *PLoS Computational Biology* 10, e1003845.

267. Maynard Smith, J. 1978. *The Evolution of Sex.* Cambridge: Cambridge University Press.

268. Maynard Smith, J. 1987. When learning guides evolution. *Nature* 329, 761-762.

269. Maynard Smith, J. & Szathmáry, E. 1995. *The Major Transitions in Evolution.* San Francisco: Freeman.

270. Mayr, E. 1982. *The Growth of Biological Thought.* Cambridge, MA: Harvard University Press.

271. McAdams, H. H. & Arkin, A. 1997. Stochastic mechanisms in gene expression. *Proceedings of the National Academy of Sciences* 94, 814-819.

272. McCommis, K. S. & Finck, B. N. 2015. Mitochondrial pyruvate transport: a historical perspective and future research directions. *Biochemical Journal* 466, 443-454.

273. McInerney, M. J. et al. 2008. Physiology, ecology, phylogeny, and genomics of microorganisms capable of syntrophic metabolism. *Annals of the New York Academy of Sciences* 1125, 58–72.

274. McNamara, J. M. 1995. Implicit frequency dependence and kin selection in fluctuating environments. *Evolutionary Ecology* 9, 185–203.

275. McRose, D. L., Baars, O., Seyedsayamdost, M. R. & Morel, F. M. 2018. Quorum sensing and iron regulate a two-for-one siderophore gene cluster in *Vibrio harveyi*. *Proceedings of the National Academy of Sciences* 115, 7581–7586.

276. Méheust, R. et al. 2021. Post-translational flavinylation is associated with diverse extracytosolic redox functionalities throughout bacterial life. *eLife* 10, e66878.

277. Meysman, F. J. 2018. Cable bacteria take a new breath using long-distance electricity. *Trends in Microbiology* 26, 411–422.

278. Millard, P. et al. 2021. Control and regulation of acetate overflow in *Escherichia coli*. *eLife* 10, e63661.

279. Molenaar, D., Van Berlo, R., De Ridder, D. & Teusink, B. 2009. Shifts in growth strategies reflect tradeoffs in cellular economics. *Molecular Systems Biology* 5, 323.

280. Moore, A. J., Brodie, E. D. & Wolf, J. B. 1997. Interacting phenotypes and the evolutionary process: I. Direct and indirect effects of social interactions. *Evolution* 51, 1352–1362.

281. Moreno, A. R. & Martiny, A. C. 2018. Ecological stoichiometry of ocean plankton. *Annual Review of Marine Science* 10, 43–69.

282. Moreno-Gámez, S. et al. 2020. Wide lag time distributions break a trade-off between reproduction and survival in bacteria. *Proceedings of the National Academy of Sciences* 117, 18729–18736.

283. Morgan, S. L. & Winship, C. 2015. *Counterfactuals and Causal Inference: Methods and Principles for Social Research*. 2nd ed. Cambridge: Cambridge Univerity Press.

284. Mori, M., Marinari, E. & De Martino, A. 2019. A yield-cost trade-off governs *Escherichia coli*'s decision between fermentation and respiration in carbon-limited growth. *NPJ Systems Biology and Applications* 5, 1–9.

285. Morin, M. et al. 2020. Genomewide stabilization of mRNA during a "feast-to-famine" growth transition in *Escherichia coli*. *mSphere* 5, e00276-20.

286. Morita, R. Y. 1997. *Bacteria in Oligotrophic Environments*. New York: Chapman & Hall.

287. Motlagh, H., Wrabl, J., Li, J. & Hilser, V. 2014. The ensemble nature of allostery. *Nature* 508, 331–339.

288. Müller, A. L. et al. 2021. An alternative resource allocation strategy in the chemolithoautotrophic archaeon *Methanococcus maripaludis*. *Proceedings of the National Academy of Sciences* 118, e2025854118.

289. Müller, H., Marozava, S., Probst, A. J. & Meckenstock, R. U. 2020. Groundwater cable bacteria conserve energy by sulfur disproportionation. *ISME Journal* 14, 623–634.

290. Nairn, J. et al. 1996. Cloning and sequencing of a gene encoding pyruvate kinase from *Schizosaccharomyces pombe:* implications for quaternary structure and regulation of the enzyme. *FEMS Microbiology Letters* 3, 329.

291. Nakaoka, H. & Wakamoto, Y. 2017. Aging, mortality, and the fast growth trade-off of *Schizosaccharomyces pombe*. *PLoS Biology* 15, e2001109.

292. New, A. M. et al. 2014. Different levels of catabolite repression optimize growth in stable and variable environments. *PLoS Biology* 12, e1001764.

293. Newman, J. R. et al. 2006. Single-cell proteomic analysis of *S. cerevisiae* reveals the architecture of biological noise. *Nature* 441, 840–846.

294. Nicholls, D. G. & Ferguson, S. J. 2013. *Bioenergetics*. 4th ed. Amsterdam: Elsevier.

295. Niehus, R. et al. 2017. The evolution of siderophore production as a competitive trait. *Evolution* 71, 1443–1455.

296. Nielsen, L. P. & Risgaard-Petersen, N. 2015. Rethinking sediment biogeochemistry after the discovery of electric currents. *Annual Review of Marine Science* 7, 425–442.

297. Nielsen, P. H. et al. 2012. Microbial communities involved in enhanced biological phosphorus removal from wastewater—a model system in environmental biotechnology. *Current Opinion in Biotechnology* 23, 452–459.

298. Nilsson, A. & Nielsen, J. 2016. Metabolic trade-offs in yeast are caused by F1F0-ATP synthase. *Scientific Reports* 6, 22264.

299. Norris, N. R. 2019. *Mechanistic Modeling of Bacterial Nutrient Uptake Strategies.* PhD thesis, Massachusetts Institute of Technology.

300. Novak, M. et al. 2006. Experimental tests for an evolutionary trade-off between growth rate and yield in *E. coli. American Naturalist* 168, 242–251.

301. Nyström, T. 2004. Growth versus maintenance: a trade-off dictated by RNA polymerase availability and sigma factor competition? *Molecular Microbiology* 54, 855–862.

302. Odling-Smee, F. J., Laland, K. N. & Feldman, M. W. 2013. *Niche Construction: The Neglected Process in Evolution.* Princeton, NJ: Princeton University Press.

303. Ogata, K. 2009. *Modern Control Engineering.* 5th ed. New York: Prentice Hall.

304. Okasha, S. 2011. Optimal choice in the face of risk: decision theory meets evolution. *Philosophy of Science* 78, 83–104.

305. Orr, H. A. 2005. The genetic theory of adaptation: a brief history. *Nature Reviews Genetics* 6, 119–127.

306. Orr, H. A. 2007. Absolute fitness, relative fitness, and utility. *Evolution* 61, 2997–3000.

307. Orth, J. D., Thiele, I. & Palsson, B. Ø. 2010. What is flux balance analysis? *Nature Biotechnology* 28, 245–248.

308. Otto, S. P. 2014. Evolution of modifier genes and biological systems. In: *The Princeton Guide to Evolution.* Ed. by J. B. Losos et al. Princeton, NJ: Princeton University Press, 253–260.

309. Park, J. O. et al. 2016. Metabolite concentrations, fluxes and free energies imply efficient enzyme usage. *Nature Chemical Biology* 12, 482–489.

310. Park, J. O. et al. 2019. Near-equilibrium glycolysis supports metabolic homeostasis and energy yield. *Nature Chemical Biology* 15, 1001–1008.

311. Payne, J. L. & Wagner, A. 2019. The causes of evolvability and their evolution. *Nature Reviews Genetics* 20, 24–38.

312. Pearl, J. & Mackenzie, D. 2018. *The Book of Why: The New Science of Cause and Effect.* New York: Basic Books.

313. Peebo, K. et al. 2015. Proteome reallocation in *Escherichia coli* with increasing specific growth rate. *Molecular BioSystems* 11, 1184–1193.

314. Pekař, M. 2020. Thermodynamic driving forces and chemical reaction fluxes; reflections on the steady state. *Molecules* 25, 699.

315. Peter, I. S. & Davidson, E. H. 2015. *Genomic Control Process: Development and Evolution.* San Diego: Academic Press.

316. Pfeiffer, T. & Morley, A. 2014. An evolutionary perspective on the Crabtree effect. *Frontiers in Molecular Biosciences* 1, 17.

317. Pfeiffer, T., Schuster, S. & Bonhoeffer, S. 2001. Cooperation and competition in the evolution of ATP-producing pathways. *Science* 292, 504-507.

318. Phan, K. & Ferenci, T. 2017. The fitness costs and trade-off shapes associated with the exclusion of nine antibiotics by OmpF porin channels. *ISME Journal* 11, 1472-1482.

319. Phillips, R. 2020. *The Molecular Switch: Signaling and Allostery.* Princeton, NJ: Princeton University Press.

320. Phillips, R. & Milo, R. 2009. A feeling for the numbers in biology. *Proceedings of the National Academy of Sciences* 106, 21465-21471.

321. Pigliucci, M. 2001. *Phenotypic Plasticity: Beyond Nature and Nurture.* Baltimore, MD: Johns Hopkins University Press.

322. Pirt, S. 1965. The maintenance energy of bacteria in growing cultures. *Proceedings of the Royal Society of London B* 163, 224-231.

323. Plata, G., Henry, C. S. & Vitkup, D. 2015. Long-term phenotypic evolution of bacteria. *Nature* 517, 369-372.

324. Porter, S. S. & Rice, K. J. 2013. Trade-offs, spatial heterogeneity, and the maintenance of microbial diversity. *Evolution* 67, 599-608.

325. Portillo, M. C., Leff, J. W., Lauber, C. L. & Fierer, N. 2013. Cell size distributions of soil bacterial and archaeal taxa. *Applied and Environmental Microbiology* 79, 7610-7617.

326. Postma, E., Verduyn, C., Scheffers, W. A. & Van Dijken, J. P. 1989. Enzymic analysis of the Crabtree effect in glucose-limited chemostat cultures of *Saccharomyces cerevisiae. Applied and Environmental Microbiology* 55, 468-477.

327. Postmus, J. et al. 2012. Isoenzyme expression changes in response to high temperature determine the metabolic regulation of increased glycolytic flux in yeast. *FEMS Yeast Research* 12, 571-581.

328. Powers, H. J. 2003. Riboflavin (vitamin B-2) and health. *American Journal of Clinical Nutrition* 77, 1352-1360.

329. Price, M. N. et al. 2013. Indirect and suboptimal control of gene expression is widespread in bacteria. *Molecular Systems Biology* 9, 660.

330. Proenca, A. M. et al. 2018. Age structure landscapes emerge from the equilibrium between aging and rejuvenation in bacterial populations. *Nature Communications* 9, 1-11.

331. Proulx, S. R. & Adler, F. R. 2010. The standard of neutrality: still flapping in the breeze? *Journal of Evolutionary Biology* 23, 1339-1350.

332. Pudlo, N. A. et al. 2015. Symbiotic human gut bacteria with variable metabolic priorities for host mucosal glycans. *mBio* 6, e01282-15.

333. Queller, D. C. 1994. Genetic relatedness in viscous populations. *Evolutionary Ecology* 8, 70-73.

334. Rabsch, W. & Winkelmann, G. 1991. The specificity of bacterial siderophore receptors probed by bioassays. *Biology of Metals* 4, 244-250.

335. Ramin, K. I. & Allison, S. D. 2019. Bacterial tradeoffs in growth rate and extracellular enzymes. *Frontiers in Microbiology* 10, 2956.

336. Ratnieks, F. L. W. 1988. Reproductive harmony via mutual policing by workers in eusocial Hymenoptera. *American Naturalist* 132, 217-236.

337. Real, L. 1980. Fitness, uncertainty, and the role of diversification in evolution and behavior. *American Naturalist* 115, 623-638.

338. Reimers, C. E. et al. 2017. The identification of cable bacteria attached to the anode of a benthic microbial fuel cell: evidence of long distance extracellular electron transport to electrodes. *Frontiers in Microbiology* 8, 2055.

339. Reintjes, G., Arnosti, C., Fuchs, B. & Amann, R. 2019. Selfish, sharing and scavenging bacteria in the Atlantic Ocean: a biogeographical study of bacterial substrate utilisation. *ISME Journal* 13, 1119-1132.

340. Ricci-Tam, C. et al. 2021. Decoupling transcription factor expression and activity enables dimmer switch gene regulation. *Science* 372, 292-295.

341. Rice, S. H. 2008. A stochastic version of the Price equation reveals the interplay of deterministic and stochastic processes in evolution. *BMC Evolutionary Biology* 8, 262.

342. Risgaard-Petersen, N., Revil, A., Meister, P. & Nielsen, L. P. 2012. Sulfur, iron-, and calcium cycling associated with natural electric currents running through marine sediment. *Geochimica et Cosmochimica Acta* 92, 1–13.

343. Roller, B. R., Stoddard, S. F. & Schmidt, T. M. 2016. Exploiting rRNA operon copy number to investigate bacterial reproductive strategies. *Nature Microbiology* 1, 1–7.

344. Ronce, O. 2007. How does it feel to be like a rolling stone? Ten questions about dispersal evolution. *Annual Review of Ecology and Systematics* 38, 231–253.

345. Rose, M. R. 1991. *Evolutionary Biology of Aging.* Oxford: Oxford University Press.

346. Roszak, D. & Colwell, R. 1987. Survival strategies of bacteria in the natural environment. *Microbiological Reviews* 51, 365–379.

347. Rotaru, A.-E. et al. 2014. A new model for electron flow during anaerobic digestion: direct interspecies electron transfer to *Methanosaeta* for the reduction of carbon dioxide to methane. *Energy & Environmental Science* 7, 408–415.

348. Russell, J. B. 2007. The energy spilling reactions of bacteria and other organisms. *Journal of Molecular Microbiology and Biotechnology* 13, 1–11.

349. Rutherford, S. L. & Lindquist, S. 1998. Hsp90 as a capacitor for morphological evolution. *Nature* 396, 336–342.

350. Samuelson, P. A. 1983. *Foundations of Economic Analysis, Enlarged ed.* Cambridge, MA: Harvard University Press.

351. Santos, J. M. et al. 2020. A novel metabolic-ASM model for full-scale biological nutrient removal systems. *Water Research* 171, 115373.

352. Santostefano, F. et al. 2020. Social selection acts on behavior and body mass but does not contribute to the total selection differential in eastern chipmunks. *Evolution* 74, 89–102.

353. Saunders, S. H. & Newman, D. K. 2018. Extracellular electron transfer transcends microbe-mineral interactions. *Cell Host & Microbe* 24, 611–613.

354. Schauer, R. et al. 2014. Succession of cable bacteria and electric currents in marine sediment. *ISME Journal* 8, 1314–1322.

355. Schavemaker, P. E., Boersma, A. J. & Poolman, B. 2018. How important is protein diffusion in prokaryotes? *Frontiers in Molecular Biosciences* 5, 93.

356. Scheiner, S. M. 1993. Genetics and evolution of phenotypic plasticity. *Annual Review of Ecology and Systematics* 24, 35–68.

357. Schink, B. 1997. Energetics of syntrophic cooperation in methanogenic degradation. *Microbiology and Molecular Biology Reviews* 61, 262–280.

358. Schlichting, C. D. & Pigliucci, M. 1998. *Phenotypic Evolution: A Reaction Norm Perspective.* Sunderland, MA: Sinauer Associates.

359. Schnell, S. & Turner, T. E. 2004. Reaction kinetics in intracellular environments with macromolecular crowding: simulations and rate laws. *Progress in Biophysics and Molecular Biology* 85, 235–260.

360. Schofield, Z. et al. 2020. Bioelectrical understanding and engineering of cell biology. *Journal of the Royal Society Interface* 17, 20200013.

361. Scholz, V. V. et al. 2021. Cable bacteria at oxygen-releasing roots of aquatic plants: a widespread and diverse plant–microbe association. *New Phytologist*, doi: 10.1111/nph.17415.

362. Schrödinger, E. 1967. *What Is Life?* Cambridge: Cambridge University Press.

363. Schuetz, R., Kuepfer, L. & Sauer, U. 2007. Systematic evaluation of objective functions for predicting intracellular fluxes in *Escherichia coli. Molecular Systems Biology* 3, 119.

364. Schuetz, R. et al. 2012. Multidimensional optimality of microbial metabolism. *Science* 336, 601–604.

365. Schulte, P. M. 2015. The effects of temperature on aerobic metabolism: towards a mechanistic understanding of the responses of ectotherms to a changing environment. *Journal of Experimental Biology* 218, 1856–1866.

366. Schuster, S., Pfeiffer, T. & Fell, D. A. 2008. Is maximization of molar yield in metabolic networks favoured by evolution? *Journal of Theoretical Biology* 252, 497–504.

367. Schut, F., Prins, R. A. & Gottschal, J. C. 1997. Oligotrophy and pelagic marine bacteria: facts and fiction. *Aquatic Microbial Ecology* 12, 177–202.

368. Scilipoti, S. et al. 2021. Oxygen consumption of individual cable bacteria. *Science Advances* 7, eabe1870.

369. Scott, M., Klumpp, S., Mateescu, E. M. & Hwa, T. 2014. Emergence of robust growth laws from optimal regulation of ribosome synthesis. *Molecular Systems Biology* 10, 747.

370. Seger, J. & Brockmann, H. J. 1987. What is bet-hedging? *Oxford Surveys in Evolutionary Biology* 4, 182–211.

371. Segre, D., Vitkup, D. & Church, G. M. 2002. Analysis of optimality in natural and perturbed metabolic networks. *Proceedings of the National Academy of Sciences* 99, 15112–15117.

372. Sekar, K. et al. 2020. Bacterial glycogen provides short-term benefits in changing environments. *Applied and Environmental Microbiology* 86.

373. Sela, I., Wolf, Y. I. & Koonin, E. V. 2016. Theory of prokaryotic genome evolution. *Proceedings of the National Academy of Sciences* 113, 11399–11407.

374. Sheridan, P. O. et al. 2016. Polysaccharide utilization loci and nutritional specialization in a dominant group of butyrate-producing human colonic Firmicutes. *Microbial Genomics* 2, e000043.

375. Shi, H., Hu, Y. & Huang, K. C. 2021. Precise regulation of the relative rates of surface area and volume synthesis in dynamic environments. *Nature Communications* 12, 1975.

376. Shi, L. et al. 2016. Extracellular electron transfer mechanisms between microorganisms and minerals. *Nature Reviews Microbiology* 14, 651–662.

377. Shin, Y. & Brangwynne, C. P. 2017. Liquid phase condensation in cell physiology and disease. *Science* 357, eaaf4382.

378. Shively, J. M., ed. 2006. *Inclusions in Prokaryotes.* Vol. 1. Berlin: Springer Science & Business Media.

379. Shoval, O. et al. 2012. Evolutionary trade-offs, Pareto optimality, and the geometry of phenotype space. *Science* 336, 1157–1160.

380. Siegal, M. L. 2015. Shifting sugars and shifting paradigms. *PLoS Biology* 13, e1002068.

381. Sies, H., Berndt, C. & Jones, D. P. 2017. Oxidative stress. *Annual Review of Biochemistry* 86, 715–748.

382. Siezen, R. J. & Hylckama Vlieg, J. E. van. 2011. Genomic diversity and versatility of *Lactobacillus plantarum*, a natural metabolic engineer. *Microbial Cell Factories* 10, S3.

383. Silva, L. G. da, Tomás-Martínez, S., Loosdrecht, M. C. van & Wahl, S. A. 2019. The environment selects: modeling energy allocation in microbial communities under dynamic environments. *bioRxiv*, doi: 10.1101/689174.

384. Simonet, C. & McNally, L. 2021. Kin selection explains the evolution of cooperation in the gut microbiota. *Proceedings of the National Academy of Sciences* 118, e2016046118.

385. Singh, R., Mailloux, R. J., Puiseux-Dao, S. & Appanna, V. D. 2007. Oxidative stress evokes a metabolic adaptation that favors increased NADPH synthesis and decreased NADH production in *Pseudomonas fluorescens*. *Journal of Bacteriology* 189, 6665-6675.

386. Slatkin, M. 1974. Hedging one's evolutionary bets. *Nature* 250, 704-705.

387. Smith, C. C. & Fretwell, S. D. 1974. The optimal balance between size and number of offspring. *American Naturalist* 108, 499-506.

388. Smith, P. & Schuster, M. 2019. Public goods and cheating in microbes. *Current Biology* 29, R442-R447.

389. Solopova, A. et al. 2014. Bet-hedging during bacterial diauxic shift. *Proceedings of the National Academy of Sciences* 111, 7427-7432.

390. Song, H.-K. et al. 2017. Bacterial strategies along nutrient and time gradients, revealed by metagenomic analysis of laboratory microcosms. *FEMS Microbiology Ecology* 93, doi: 10.1093/femsec/fix114.

391. Sorensen, J., Dunivin, T., Tobin, T. & Shade, A. 2019. Ecological selection for small microbial genomes along a temperate-to-thermal soil gradient. *Nature Microbiology* 4, 55-61.

392. Stearns, S. C. 1992. *The Evolution of Life Histories*. New York: Oxford University Press.

393. Stearns, S. C. 2000. Life history evolution: successes, limitations, and prospects. *Naturwissenschaften* 87, 476-486.

394. Stincone, A. et al. 2015. The return of metabolism: biochemistry and physiology of the pentose phosphate pathway. *Biological Reviews* 90, 927-963.

395. Stocker, R. 2012. Marine microbes see a sea of gradients. *Science* 338, 628-633.

396. Storz, G. & Imlayt, J. A. 1999. Oxidative stress. *Current Opinion in Microbiology* 2, 188-194.

397. Strassmann, J. E. & Queller, D. C. 2011. Evolution of cooperation and control of cheating in a social microbe. *Proceedings of the National Academy of Sciences* 108, 10855-10862.

398. Surovtsev, I. V. & Jacobs-Wagner, C. 2018. Subcellular organization: a critical feature of bacterial cell replication. *Cell* 172, 1271-1293.

399. Szenk, M., Dill, K. A. & Graff, A. M. de. 2017. Why do fast-growing bacteria enter overflow metabolism? Testing the membrane real estate hypothesis. *Cell Systems* 5, 95-104.

400. Tal, O. & Tran, T. D. 2020. Adaptive bet-hedging revisited: considerations of risk and time horizon. *Bulletin of Mathematical Biology* 82, 50.

401. Tatenhove-Pel, R. J. van et al. 2021. Serial propagation in water-in-oil emulsions selects for *Saccharomyces cerevisiae* strains with a reduced cell size or an increased biomass yield on glucose. *Metabolic Engineering* 64, 1-14.

402. Taylor, J. R. & Stocker, R. 2012. Trade-offs of chemotactic foraging in turbulent water. *Science* 338, 675-679.

403. Taylor, P. D. 1990. Allele-frequency change in a class-structured population. *American Naturalist* 135, 95-106.

404. Taylor, P. D. 1992. Altruism in viscous populations—an inclusive fitness approach. *Evolutionary Ecology* 6, 352-356.

405. Taylor, P. D. & Frank, S. A. 1996. How to make a kin selection model. *Journal of Theoretical Biology* 180, 27-37.

406. Teusink, B. et al. 2009. Understanding the adaptive growth strategy of *Lactobacillus plantarum* by in silico optimisation. *PLoS Computational Biology* 5, e1000410.

407. Thauer, R. K. et al. 2008. Methanogenic archaea: ecologically relevant differences in energy conservation. *Nature Reviews Microbiology* 6, 579-591.

408. *The Economist.* The Japanification of bond markets. August 22, 2019.

409. Thrash, J. C. & Coates, J. D. 2008. Direct and indirect electrical stimulation of microbial metabolism. *Environmental Science & Technology* 42, 3921-3931.

410. Thucydides. 1914. *History of the Peloponnesian War Done into English by Richard Crawley.* London: J. M. Dent & Sons, Ltd.

411. Travisano, M. & Velicer, G. J. 2004. Strategies of microbial cheater control. *Trends in Microbiology* 12, 72-78.

412. Tsai, C.-J. & Nussinov, R. 2014. A unified view of "how allostery works." *PLoS Computational Biology* 10, e1003394.

413. Tseng, C.-P., Albrecht, J. & Gunsalus, R. P. 1996. Effect of microaerophilic cell growth conditions on expression of the aerobic (cyoABCDE and cydAB) and anaerobic (narGHJI, frdABCD, and dmsABC) respiratory pathway genes in *Escherichia coli. Journal of Bacteriology* 178, 1094-1098.

414. Tuljapurkar, S., Gaillard, J. M. & Coulson, T. 2009. From stochastic environments to life histories and back. *Philosophical Transactions of the Royal Society B* 364, 1499.

415. Turner, J. J., Ewald, J. C. & Skotheim, J. M. 2012. Cell size control in yeast. *Current Biology* 22, R350-R359.

416. Ude, J. et al. 2021. Outer membrane permeability: antimicrobials and diverse nutrients bypass porins in *Pseudomonas aeruginosa. Proceedings of the National Academy of Sciences* 118, e21076-44118.

417. Vadia, S. & Levin, P. A. 2015. Growth rate and cell size: a reexamination of the growth law. *Current Opinion in Microbiology* 24, 96-103.

418. Valgepea, K., Adamberg, K., Seiman, A. & Vilu, R. 2013. *Escherichia coli* achieves faster growth by increasing catalytic and translation rates of proteins. *Molecular BioSystems* 9, 2344-2358.

419. Valgepea, K., Adamberg, K. & Vilu, R. 2011. Decrease of energy spilling in *Escherichia coli* continuous cultures with rising specific growth rate and carbon wasting. *BMC Systems Biology* 5, 1-11.

420. Valko, M. et al. 2007. Free radicals and antioxidants in normal physiological functions and human disease. *International Journal of Biochemistry & Cell Biology* 39, 44-84.

421. Van den Bergh, B., Fauvart, M. & Michiels, J. 2017. Formation, physiology, ecology, evolution and clinical importance of bacterial persisters. *FEMS Microbiology Reviews* 41, 219-251.

422. Vedel, S. et al. 2016. Asymmetric damage segregation constitutes an emergent population-level stress response. *Cell Systems* 3, 187-198.

423. Vemuri, G. N. et al. 2006. Overflow metabolism in *Escherichia coli* during steady-state growth: transcriptional regulation and effect of the redox ratio. *Applied and Environmental Microbiology* 72, 3653-3661.

424. Vemuri, G. N. et al. 2007. Increasing NADH oxidation reduces overflow metabolism in *Saccharomyces cerevisiae*. *Proceedings of the National Academy of Sciences* 104, 2402-2407.

425. Venturelli, O. S., Zuleta, I., Murray, R. M. & El-Samad, H. 2015. Population diversification in a yeast metabolic program promotes anticipation of environmental shifts. *PLoS Biology* 13, e1002042.

426. Vinnicombe, G. 2001. *Uncertainty and Feedback: H_∞ Loop-Shaping and the ν-Gap Metric*. London: Imperial College Press.

427. Visser, J. de et al. 2003. Perspective: evolution and detection of genetic robustness. *Evolution* 57, 1959-1972.

428. Voet, D., Voet, J. G. & Pratt, C. W. 2011. *Biochemistry*. 4th ed. Hoboken, NJ: Wiley.

429. Voit, E. O. 2000. *Computational Analysis of Biochemical Systems: A Practical Guide for Biochemists and Molecular Biologists*. Cambridge: Cambridge University Press.

430. Volkmer, B. & Heinemann, M. 2011. Condition-dependent cell volume and concentration of *Escherichia coli* to facilitate data conversion for systems biology modeling. *PloS ONE* 6, e23126.

431. Waddington, C. H. 1942. Canalization of development and the inheritance of acquired characters. *Nature* 154, 563-565.

432. Waddington, C. H. 1953. Genetic assimilation of an acquired character. *Evolution* 7, 118-126.

433. Wade, M. J. 1978. A critical review of the models of group selection. *Quarterly Review of Biology* 53, 101-114.

434. Wade, M. J. 1985. Soft selection, hard selection, kin selection, and group selection. *American Naturalist* 125, 61-73.

435. Wagner, A. 2013. *Robustness and Evolvability in Living Systems*. Princeton, NJ: Princeton University Press.

436. Wagner, G. P. & Altenberg, L. 1996. Complex adaptations and the evolution of evolvability. *Evolution* 50, 967-976.

437. Wang, J. et al. 2015. Natural variation in preparation for nutrient depletion reveals a cost–benefit tradeoff. *PLoS Biology* 13, e1002041.

438. Wang, L. et al. 2020. Recent progress in the structure of glycogen serving as a durable energy reserve in bacteria. *World Journal of Microbiology and Biotechnology* 36, 1-12.

439. Wechsler, T., Kümmerli, R. & Dobay, A. 2019. Understanding policing as a mechanism of cheater control in cooperating bacteria. *Journal of Evolutionary Biology* 32, 412–424.

440. Weiße, A. Y., Oyarzún, D. A., Danos, V. & Swain, P. S. 2015. Mechanistic links between cellular trade-offs, gene expression, and growth. *Proceedings of the National Academy of Sciences* 112, E1038–E1047.

441. West, S. A. et al. 2007. The social lives of microbes. *Annual Review of Ecology and Systematics* 38, 53–77.

442. West-Eberhard, M. J. 1983. Sexual selection, social competition, and speciation. *Quarterly Review of Biology* 58, 155–183.

443. West-Eberhard, M. J. 2003. *Developmental Plasticity and Evolution.* New York: Oxford University Press.

444. Westerhoff, H. V., Hellingwerf, K. J. & Van Dam, K. 1983. Thermodynamic efficiency of microbial growth is low but optimal for maximal growth rate. *Proceedings of the National Academy of Sciences* 80, 305–309.

445. Westfall, C. S. & Levin, P. A. 2017. Bacterial cell size: multifactorial and multifaceted. *Annual Review of Microbiology* 71, 499–517.

446. Williams, G. C. 1966. *Adaptation and Natural Selection.* Princeton, NJ: Princeton University Press.

447. Wilson, D. S. 1980. *The Natural Selection of Populations and Communities.* Menlo Park, CA: Benjamin/Cummings.

448. Wilson, D. S. & Dugatkin, L. A. 1997. Group selection and assortative interactions. *American Naturalist* 149, 336–351.

449. Wilson, D. S., Pollock, G. B. & Dugatkin, L. A. 1992. Can altruism evolve in purely viscous populations? *Evolutionary Ecology* 6, 331–341.

450. Wilson, W. A. et al. 2010. Regulation of glycogen metabolism in yeast and bacteria. *FEMS Microbiology Reviews* 34, 952–985.

451. Windels, E. M. et al. 2019. Bacterial persistence promotes the evolution of antibiotic resistance by increasing survival and mutation rates. *ISME Journal* 13, 1239–1251.

452. Wolfe, A. J. 2005. The acetate switch. *Microbiology and Molecular Biology Reviews* 69, 12–50.

453. Wong, W. W., Tran, L. M. & Liao, J. C. 2009. A hidden square-root boundary between growth rate and biomass yield. *Biotechnology and Bioengineering* 102, 73–80.

454. Wortel, M. T. et al. 2018. Metabolic enzyme cost explains variable trade-offs between microbial growth rate and yield. *PLoS Computational Biology* 14, e1006010.

455. Wright, S. 1931. Evolution in Mendelian populations. *Genetics* 16, 97-159.

456. Wright, S. 1932. The roles of mutation, inbreeding, cross-breeding and selection in evolution. *Proceedings VI International Congress of Genetics* 1, 356-366.

457. Wu, M. et al. 2015. Genetic determinants of in vivo fitness and diet responsiveness in multiple human gut *Bacteroides*. *Science* 350, aac5992.

458. Xiao, X. & Yu, H.-Q. 2020. Molecular mechanisms of microbial transmembrane electron transfer of electrochemically active bacteria. *Current Opinion in Chemical Biology* 59, 104-110.

459. Xu, J. et al. 2020. Metabolic flux analysis and fluxomics-driven determination of reaction free energy using multiple isotopes. *Current Opinion in Biotechnology* 64, 151-160.

460. Yamamura, N., Higashi, M., Behera, N. & Wakano, J. Y. 2004. Evolution of mutualism through spatial effects. *Journal of Theoretical Biology* 226, 421-428.

461. Yang, Y. et al. 2019. Temporal scaling of aging as an adaptive strategy of *Escherichia coli*. *Science Advances* 5, eaaw2069.

462. Yin, H., Aller, R. C., Zhu, Q. & Aller, J. Y. 2021. The dynamics of cable bacteria colonization in surface sediments: a 2D view. *Scientific Reports* 11, 1-8.

463. Yooseph, S. et al. 2010. Genomic and functional adaptation in surface ocean planktonic prokaryotes. *Nature* 468, 60-66.

464. You, C. et al. 2013. Coordination of bacterial proteome with metabolism by cyclic AMP signalling. *Nature* 500, 301-306.

465. Youk, H. & Van Oudenaarden, A. 2009. Growth landscape formed by perception and import of glucose in yeast. *Nature* 462, 875-879.

466. Zakrzewska, A. et al. 2011. Genome-wide analysis of yeast stress survival and tolerance acquisition to analyze the central trade-off between growth rate and cellular robustness. *Molecular Biology of the Cell* 22, 4435-4446.

467. Zerfaß, C., Asally, M. & Soyer, O. S. 2019. Interrogating metabolism as an electron flow system. *Current Opinion in Systems Biology* 13, 59–67.

468. Zerfaß, C., Chen, J. & Soyer, O. S. 2018. Engineering microbial communities using thermodynamic principles and electrical interfaces. *Current Opinion in Biotechnology* 50, 121–127.

469. Zhao, C. et al. 2017. Reexamination of the physiological role of PykA in *Escherichia coli* revealed that it negatively regulates the intracellular ATP levels under anaerobic conditions. *Applied and Environmental Microbiology* 83, e00316–17.

470. Zhong, S., Miller, S. P., Dykhuizen, D. E. & Dean, A. M. 2009. Transcription, translation, and the evolution of specialists and generalists. *Molecular Biology and Evolution* 26, 2661–2678.

471. Zhou, K. & Doyle, J. C. 1998. *Essentials of Robust Control.* Upper Saddle River, NJ: Prentice Hall.

472. Zhuang, K., Vemuri, G. N. & Mahadevan, R. 2011. Economics of membrane occupancy and respiro-fermentation. *Molecular Systems Biology* 7, 500.

473. Zimmerman, A. E., Martiny, A. C. & Allison, S. D. 2013. Microdiversity of extracellular enzyme genes among sequenced prokaryotic genomes. *ISME Journal* 7, 1187–1199.

474. Zuo, W. & Wu, Y. 2020. Dynamic motility selection drives population segregation in a bacterial swarm. *Proceedings of the National Academy of Sciences* 117, 4693–4700.

Index

CPSIA information can be obtained
at www.ICGtesting.com
Printed in the USA
JSHW040247230622
27203JS00001B/1

9 780691 231198